WILD
BRITAIN
THE CENTURY BOOK OF
MARSHES, FENS & BROADS

WILD
BRITAIN

THE CENTURY BOOK OF
MARSHES, FENS & BROADS

Richard North
Foreword by David Bellamy

CENTURY PUBLISHING
LONDON

Produced by Robert Dudley and John Stidolph
Antler Books Ltd
11 Rathbone Place
London W1P 1DE

Maps by Martin Lubikowski
Designed by Ian Hughes
Edited by Wendy Slemen

First published in Great Britain in 1983 by
Century Publishing Co Ltd,
76 Old Compton St, London W1V 5PA
ISBN 0 7126 0195 3

Typeset by TJB Photosetting Ltd
South Witham, Lincolnshire
Pictures originated by Springbourne Press Ltd,
Basildon, Essex
Printed and bound in Italy
by New Interlitho SpA Milan

Contents

Acknowledgements
page vii

Foreword by David Bellamy
page ix

CHAPTER ONE

*A Streamside:
the start of it all*

page 10

CHAPTER TWO

Norfolk's Broadland

page 22

CHAPTER THREE

*The
Somerset Levels*

page 54

CHAPTER FOUR

*The East Anglian
Fenlands*

page 74

CHAPTER FIVE

*Freshwater Marshes
and Wet Meadows*

page 90

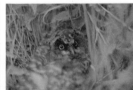

CHAPTER SIX

The Urban Wetlands

page 118

CHAPTER SEVEN

The Acid Bogs

page 128

CHAPTER EIGHT

*Estuaries and
Saltmarshes*

page 162

CHAPTER NINE

*The
Conservation Case*

page 178

Booklist
page 190

Useful Addresses
page 191

picture acknowledgements
page 192

Acknowledgements

This book would never have been written without the unstinting generosity of several people who have devoted themselves to understanding and defending wetlands, and who gave very freely of their time to help me.

To David Brewster of the Broads Authority, Dr Pat Doody of the Nature Conservancy Council, Dr Martin George of the Nature Conservancy Council, Dr John Harvey of Cambridge University, Jeremy Purseglove of Severn Trent Water Authority, Dr Derek Wells of the Nature Conservancy Council, Gwyn Williams of the Royal Society for the Protection of Birds, Rob Williams of the Nature Conservancy Council, John Riggall and Fiona Burd of the Nature Conservancy Council, I owe the benefit of several conversations and much advice.

I would like to thank Richard Lindsay of the Nature Conservancy Council very particularly for taking such a strong and kindly interest in the project.

However, though I asked all these people and more for advice, I did not always take it: any mistakes and inaccuracies – and certainly the style of the enterprise – are wholly my own. It was part of the generosity of those who helped me that they allowed me to make what use I liked of the material they gave me.

To Richard Lindsay, Charlie Pye-Smith, Paul Wymer, and Glyn Satterley I owe the fun of travelling together in the autumn of 1982. To Paul Wymer I owe an introduction to A G Tansley's work, which is irreplaceable. To Glyn Satterley I owe the pictures which will ensure, I think, that every page of this book on which they appear will be pored over.

Thanks are due to Philips for the continued loan of a Philips 2000 word processor, without which I could not have written the book so quickly, nor peppered my correspondents so easily with drafts of various chapters for their comments.

Finally, I owe an enormous amount to Anne, my wife, who has given me three lovely children, and kept them out of my hair while bringing them up so well that their few interruptions of the work were always welcome.

Doxey Marshes, near Stafford. A beautiful wetland near to a city environment: a perfect example of how wet-footedness can preserve wilderness near large populations.

Northern Marsh Orchid. A classic damp-loving flower, which is fairly common in the north country and Scotland.

Foreword by David Bellamy

I have for the past twenty-five years of my life spent as much time as possible immersed in the wetlands of the world and especially of the British Isles. My aim has been to gain an understanding of the relationship which exists between habitat and the plants and animals which make the wetlands one of the richest parts of our natural heritage.

When I began my studies there was still much to be discovered even in the field – a wealth of sites of all types from the poorest acid bog to the richest fen, ponds, rivers, streams still in their semi-natural state, all uncatalogued let alone studied in the detail they deserved.

This is no longer true. In those few years the British landscape has changed in places beyond recognition and worst hit of all have been the wetland sites. The vast majority of these changes have been detrimental to the wildlife and the wildflowers of these islands and to the long term stability and economy of our world-renowned and envied rural landscape, which has in the past kept both our soils and our society in good heart.

The twentieth century inherited the British landscape with all its rich diversity of habitat and custom as a legacy from a much harsher past when men, women, and children toiled in field and factory to scrape a meagre living. They were, however, fortunate on two counts. They had hope for a future and they had contact with the beauty and excitement of a living landscape. A world of topsy-turvy order bursting with butterflies and song birds, criss-crossed with hedgerows each overflowing with wildflowers, damp secret places, ducks on the village pond, daisies in the outfield, partridge, pheasant, fishing, new potatoes, tiddlers in a jam-jar, heather, honey still for tea, rights of common, each one in their own due season.

What legacy are we going to hand on to the next century and to our children's children?

Please read this book, for it concerns our wetlands, the very heart of that legacy. It traces the falling of raindrops which breathe hope into this dry earth as they join to form the life blood of our living landscapes; ebullient mountain streams, sought out each year by spawning fish which are drawn from the vastness of the oceans to clear oxygenated waters which provide their only hope for any future; the once vast morass of living upland peat which blankets the wetter west and the gentle mountains of Scotland, England and Wales. This is a living history book which not only records the passing of each year within its peaty mass but supports a diversity of life on the meagre supplies of nutrient brought in by rain. We have so much to learn from nature if only we take time to sit and watch and think. The English lakes, the Anglian broads and fens are both five star attractions which bring in tourist millions each and every year. So also do those literary landscapes, the Shakespeare, Bronte, Dickens, Hardy, Thomas, and so many more counties, each of which has a river at its heart, living water, which weaves throughout their stories as it does throughout their landscapes.

Please read this book. It tells of things past and things present and it presents two faces for the future. The decision as to which of those faces we will hand on to the twenty-first century – our legacy for the future – is in your hands.

DAVID BELLAMY

BEDBURN 1983

CHAPTER ONE

A Streamside: the start of it all

The Halladale River runs out to the sea at the lovely little village of Melvich, west of Thurso, up in the north east of Scotland. The main river is very fine, and where it meets the sea there are dramatic sand-dunes and fine walking. We – a gaggle of friends in search of the British wetlands – came across it after several weeks' exploration of the soggy parts of the British land surface. Among us, and our guide for many of the Scottish legs of the trip, was an enthusiastic lover of the wetland habitat, Richard Lindsay, of the Nature Conservancy Council (see Chapter 7).

I had suggested to Richard that a well-splashed streamside, perhaps with a water-fall, might make a useful and arresting

introduction to wetlands. I had in mind some exquisitely rich and eccentric places I had seen in the Pyrenees in high summer, where waterfalls in hot little gorges make a kind of vertical wetland (most wetlands are complete-ly horizontal). The constant drenching and the sunshine produce an effect not unlike an expensive florist's shop; a heady dank world in which ferns, carpets of moss, and great waving grass-clumps arching out over space all thrive. There was no chance of anything like that in a Scottish scene preparing itself for winter, but surely a stream could tell part of our story. Richard had hummed and

The Wetlands of Britain. Some of the most important of the country's soggy places.

1. Blar nam Faoileag: blanket bog
2. Silver Flowe: blanket bog
3. Kilhern Moss: blanket bog
4. Claish Moss: blanket bog
5. Rannoch Moor: blanket bog
6. Strathy River Bogs: blanket bog
7. Glasson Moss: raised bog
8. Leighton Moss: sedge fen
9. North Fen, Esthwaite: fen
10. Irthinghead Mires: blanket bog
11. New Forest Valley Mires: blanket bog
12. Thursley Bog: valley bog
13. Woodwalton Fen: sedge fen
14. Wicken Fen: sedge fen
15. Ouse Washes: man-made washlands
16. The Broads
17. Chippenham Fen: wooded fen
18. Somerset Levels
19. Roydon Common: valley bog
20. Western Cleddau: wet meadow and fen
21. Dowrog Common: fen and wet heath
22. Insh Marshes: sedge fen and wet meadow
23. Amberley Wild Brooks: wet meadow
24. North Meadow, Cricklade: wet meadow
25. Lower Woodford Water Meadows: water meadows
26. Derwent Ings: wet meadows
27. North Kent Marshes: drained marsh
28. Stafford Marshes: marsh and wet meadow
29. Chichester Harbour: estuarine mudflats and marshes
30. Ribble Estuary: estuarine mudflats and marshes
31. Morecambe Bay: estuarine mudflats and marshes
32. North Norfolk Coast: estuarine mudflats and marshes
33. Cromarty Firth: estuarine mudflats and marshes
34. Moray Firth: estuarine mudflats and marshes

35. Solway Firth: estuarine mudflats and marshes
36. The Humber Estuary: estuarine mudflats and marshes
37. Severn Estuary: estuarine mudflats and marshes
38. The Wash: estuarine mudflats and marshes
39. The Dee Estuary: estuarine mudflats and marshes
40. The Thames Estuary: estuarine mudflats and marshes
41. Carmarthen Bay: estuarine mudflats and marshes
42. Tees-side: estuarine mudflats and marshes
43. The Alde Estuary: estuarine mudflats and marshes
44. Hebridean Machair: rare shell-sand wet meadow formation
45. Minsmere: man-influenced sedge-fen and meadow
46. Mound Alderwoods: rare man-influenced fen-carr
47. Walthamstow Marshes: derelict wet meadow
48. Moseley Bog, Birmingham: wet woodland
49. Crymlyn Bog, Swansea: sedge fen
50. Borth Bog: raised bog
51. Tregaron Bog: raised bog
52. Oxwich: sedge fen
53. Chartley Moss: rare 'basin' bog (possibly formed over erstwhile salt dome)
54. Tarn Fen: wooded fen
55. Fen Bogs: valley bog
56. Whitlaw Mosses: sedge/moss fen
57. Morden Bog: valley bog
58. East Flanders Moss: raised bog
59. Abernethy Forest Mires: forest valley bogs
60. Inverpolly Valley Mires: valley bogs
61. Oykell Marshes: freshwater marsh and wet meadows
62. Ken-Dee Marshes: wet meadow and marsh
63. Moor House: blanket bog
64. Potteric Carr, Doncaster: urban wetland

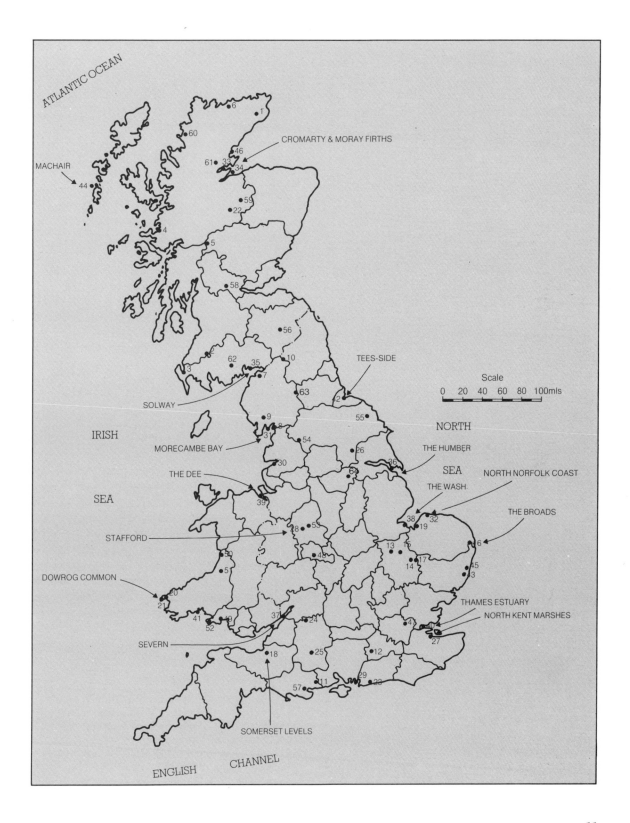

hawed for three or four days about where such a spot might occur on our planned trip, and we had all but forgotten about it.

The day we were at Melvich began as several of our days together did. I wrote a letter to the bank explaining why I needed to borrow so much more than my overdraft allowed (travelling was wonderful fun, but expensive) and then went for a walk. Richard Lindsay, full of energy, but not the earliest riser in the world, had surfaced by the time I got back. Charlie Pye-Smith, another of the ecologically-minded people without whom this book would never have been written, and another great travelling companion, was still firmly ensconced in his sleeping bag in the motorhome we had borrowed. After breakfast we headed through the great wildernesses of Caithness to see evidence of doomed 'improvements' to the peatbogs there (see Chapter 7). We were motoring down the Strath Halladale road, one of the hundreds upon hundreds of sweetly sinuous roads in the Highlands where there is no point in rushing and where passing-places alone allow cars to overtake.

Suddenly, Richard Lindsay knew exactly what had escaped his mind before. There was, he said, a little burn he had discovered with other members of the NCC's peatland team. It was a secret, rushing affair which he remembered for its extraordinary prettiness and because there was a millhouse at its meeting with the road we were on. It had other merits, he said, but did not elaborate. And so it was that at around elevenses time (when Charlie Pye-Smith could be prised from his rest) on that sunny, clear day, we parked the motor, put on waders and wellies, and set off.

The Smigel Burn hustles down a steep small gorge. You can probably march the high banks on either side. We did not find out about the topsides of the first stretch, but chose instead to try to stomp straight up the watercourse itself. The Smigel is, for much

of its course, almost the exact opposite of the wetland habitats we had set out to see. It is steep and rocky. On its immediate 'shores' of rock and boulders no very complex plant-forms could get a living, but in some ways the bouldered edge of the Smigel is the ideal place to begin an exploration of the wetlands.

Any habitat with a serious nature conservation interest is to some extent, among its other values, a museum of the progress life has made on the planet. In certain respects the wetland habitat is among the most primitive and 'natural' habitats we still have. As we shall see, the wetlands have been exploited by man in one way or another for thousands of years, but there remain pockets where they are as they always were. Often enough, even in these so-called crowded islands, when we are in wetlands we are in the presence of something immensely complex and rich which has been doing its own thing in pretty much the same way since the end of the last glaciation (around 10,000 years ago). Even this period of northern refrigeration merely represented an interruption in a process which these places had 'discovered' millions of years before.

Saltmarshes at Sheppey, on the Thames estuary. An attractive and plausible theory has it that estuaries may well have been the in-between world which gave sea-life the chance to acclimatise to the land. The wetlands themselves probably represent the most present reminder of life's aquatic beginnings.

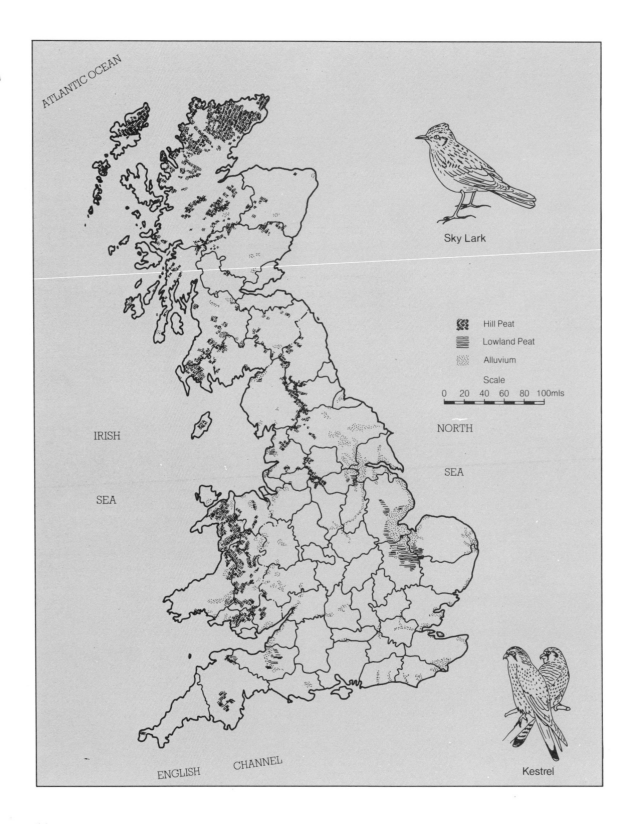

ATLANTIC OCEAN

Sky Lark

Hill Peat
Lowland Peat
Alluvium

Scale
0 20 40 60 80 100mls

IRISH

NORTH

SEA

SEA

Kestrel

ENGLISH CHANNEL

For such places to remain undisturbed they must either seem wholly useless to man, or provide in their natural state a better living for him than they would if large amounts of money and effort were spent on their reclamation. The wetlands were in part both of these until very recent historical times. The sort of exploitation that changed the wetlands' appearance did not begin in earnest until well within the time of written records: say, with the Romans (see Chapter 4). It was patchy, and it has not even now gone so far that there are not some places where the wetland remains triumphantly itself, and many others where it obstinately persists in defying change, continuing to delight us with its capacity to throw up occasional unexpected, unprofitable life-forms. Those places are the subject of this book.

For present purposes, the term 'wetland' is more or less reserved for areas which are wet but not permanently inundated to any great depth. It may not be a strict ecological definition, and it is more limited than some, but, roughly speaking, we have included those habitats in which one can walk about in gum boots and left out those which require waders or a boat. No habitats where it is possible to walk in ordinary shoes in winter have been thought wet enough, though the fringes of some would be perfectly manageable in plimsolls in summer. The term 'wetland' has been stretched beyond 'soggy land' to

include shallow patches of water. It does not include the properly aquatic environments of deep lakes, rivers, and streams. We have included the Norfolk Broads, because though they include much open water, it is very shallow water on habitat which would be very different – indeed, wholly within the book's definition – had it not been for man, and it is water with an immensely important wetland fringe. For similar reasons, farm ponds, some river fringes, and some ditches of various kinds are included. However, as with most natural-history subjects, it is extremely difficult to draw precise distinctions, and one can sometimes strain credulity in the attempt. Most of the water described in this book does not flow very fast or deep. Nevertheless, to convey a sense of the origins and the antiquity of the plant-forms and systems of which the modern wetlands are the present representatives, the Smigel Burn with its rushing waters is even better than a more typical wetland site.

For the past hundred years, there has been a good deal too much pretended certainty about how life-forms developed on the planet. This is not the place to join the debate about just how much evidence there really is for the small print of the evolutionary theories which have accreted around the great Charles Darwin's *Origin of Species*. Enough to say that there is very little fossil evidence to substantiate any description of the origin of plant species on the earth. The field is wide open for conjecture. Man will go a very long way in deceiving himself that he understands and has evidence for the rationales about the development of life-forms he is naturally keen to discuss. He does not like to admit ignorance, especially about his past, and this has led to some powerful stretching of Darwin's case, a case which the great man himself dithered over for twenty years before risking publication, perhaps being unsure whether he, too, had not overstated some of the minutiae in his theory of evolution.

Wetland terrain in Britain. These are areas where geology and climate have combined to produce damp-footed flora and fauna. Alluvium is usually the deposit of soil from rivers where they slow down and lack the energy to carry silt along: it therefore defines the kind of slow-to-drain neighbourhoods where wetlands can form. Hill peat is usually acid bog, though the vast majority of its habitat no longer supports living mosses. Lowland peat is usually fen peat, and again little of it is now growing as fen.

The tumbling stream of Smigel, boisterous and well developed though it is, is important here. On its boulders and rocks there is a fine crop of the mosses, lichens, liverworts, and ferns which (with horsetails, not found here) make up the transitional plant world which probably bridged the gulf between the plant-life of total immersion in the primordial waters and the modern land-bound flora we know today. Some of the most primitive of the plants we shall be dealing with are a group of mosses whose speciality is to live in very still water which the mosses themselves store all around them. They are plants which, in bulk and in community, provide the environment in which the individuals live: they are in that sense intensely sociable.

Mosses are primitive plants, in the sense that they have neither roots nor an internal structure with which to move water and nutrients upward from any soil beneath them. They do not have the means to probe down into the richness of water and nutrient in soil. In the case of the sphagnum (acid-bog-forming) mosses, their entire existence depends on living in nutritionally poor conditions in which they receive their nutrients solely from water. Some mosses, not even having water-tight coatings on their leaves, need constant wetness in their surroundings, for without this they would desiccate and die. All mosses need water; oddly enough the purest bog-forming mosses (see Chapter 7) – pure in the sense of being very demanding as to the conditions they require for survival – betray very primitive qualities though living in very highly developed communities which depend on a complex system of development. But some species – not the bog-formers – were perhaps capable of colonizing the dangerous fringe between land and water – something which, it is postulated, must have been essential for any plants bridging the gap between the fertile and teeming waters of 1,000 million years ago and the land-based life-forms which

were probably in evidence by 700 million years ago.

The mosses were not alone in the early world. The clubmosses, liverworts, lichens, horsetails, and ferns, many of which are still in evidence on the wetland sites under discussion, also had the sorts of qualities which would have made them capable of bridge-building. The evidence appears to suggest that 1,000 million years ago, the tropical seas were rich in plant-life which had developed the capacity to photosynthesize energy from the sun's rays. The theory goes that the sea produced the progenitors of the forms of plant-life which later found ways of living away from the sea, in environments which were increasingly dry. Seed-producing plants and – presumably later – flowering plants liberated vegetation from the need to live and reproduce in water. The lichens, liverworts, mosses, clubmosses, horsetails, and ferns still bear traces of their distant, aquatic origins during the critical phase of reproduction, when exchange of genetic material is only possible in the presence of water. Nevertheless, the last three are all capable of growing to immense size.

This is a fascinating composite infra-red picture built up from 70 million picture elements taken by the Landsat satellite. The picture understates the quantity of arable farmland, but shows peat and heather moor, and rough grazing grassland beautifully. Deep blue: sea water. Very deep blue/black: rivers, lakes. Very dark brown: woodland. Dark brown/black: heather moor. Mid-browns (greenish brown to red-brown): peat moor. Red: agricultural land. Green to cyan (green-blue): broken grassland. Blue (inland): built-up areas. Blue (coastal): sandbanks. White: cloud and snow.

To begin, then, with the most limited, and therefore presumably earlier and 'lower' plant groups – though hierarchies in which relative antiquity and simplicity, or relative modernity and complexity, are always grouped together, may themselves be a horrible over-simplification.

Lichens consist of a liaison between alga and fungus. Since algae seem likely to have been the great colonizers of the sea habitat, it is an attractive hypothesis that lichens represent the next, land-living, forms. Many lichens manage well enough, in their slow-growing way, in dry environments, but there are wet-living types of lichens too, and they do their slow thing rather faster. The fungal component of a lichen has the effect of protecting its algal host from drying-out in air; it makes a skin. The lichens living on the rocksides of the Smigel Burn show that lichens can live in the barest of environments – bare rock and stream-spray. The formation of a waterproof surface, allowing moisture to be held within (and kept out of) a plant, marks a crucial stage in the escalation of sophistication in plant-life. All these plants can reproduce by the production of spores, and their female and male cells depend on water as the medium of their meeting. The liverworts have similar limitations of lifestyle but do not congregate together as mosses do. The secret of the power of the mosses is their capacity to form huge communities which create special environments of their own, uniquely vegetative, kind (see Chapter 7).

Lichens, liverworts, and mosses do not have the woody stems which support tall growth, neither do they have the roots or the vascular ('containing vessels') systems by which nutrients and water are conducted through bigger plants. They form carpets, rather than standing up on their own. They are thus liable to be shaded out by any plant which can rise above them.

There remains an important category of primitive plants which, like lichens, liver-worts, and mosses, produces spores rather than seeds, but is still capable of great size. The ferns, horsetails, and clubmosses (and the rather peculiar, small group of quillworts) all have roots, nutrient and water 'transport' systems, and (except quillworts) the sorts of stems which support tall growth. They are a very ancient group of plants, and seem to fit neatly into the hypothetical colonization of the land by plants. The seas, as we have said, were rich in algal life, including seaweeds – which are no more than potentially rather large, floppy forms of algal growth.

The earliest real land plant of which we have a fossil record is the *Cooksonia*, which seems to have lived 400 million years ago. It has what looks like a relative alive today, in the swamp forests of Borneo. This is *Psilotum nudum*, which grows as an epiphyte on trees; this means that it depends on a host plant for its physical environment, but is not parasitic on it. It has seaweed-like branch structures. The Pteridophytes – a plant group of which *Cooksonia* was a part, and which *Psilotum nudum* and the clubmosses, horsetails, quillworts, and ferns are now a part – are vascular (they have an internal means of conducting nutrients and water) and strong-stemmed. They also seem to have bridged the water world and the land world by producing two very different sorts of generations.

Each of these plants can produce a generation which creates spores but which lives as a 'modern', sophisticated plant: it can find water in the soil beneath it, even if there is none on the surface, and it can grow tall, supported on its woody stem, which it is tempting to see as some sort of stiffened version of a seaweed stem. True, its spores must reproduce in water and through the medium of water. They produce at first a kind of plant which cannot grow tall, and only after that generation will another come which grows mighty again. But the Pteridophytes, for all that their reproduction system remained aquatic, were at least capable of a land life. They could seek water at their feet rather than needing inundation of their whole length, and they could transport water and nutrients over all their sometimes immense structures.

There are Pteridophytes in the hustling little stream of the Smigel. There is the broad, dark-green leaf of the hart's tongue fern, the pale-green hair-comb of the common polypody and the soft bottle-brush of the marsh horsetail. These are of exactly the same plant-forms that laid down the coal deposits on which we now depend. They are exactly the sort of plants which made the nineteenth-century industrial revolution possible and which we still use, perhaps too much and in a way which may threaten us. We are burning the corpses of plants of this kind, since the forests in which they grew became coal. Burning releases the carbon of which their bodies are formed, and represents their final decay as heat and carbon dioxide. This may crucially alter the structure of the gaseous outer limits of our planet and its life-support systems.

It was the peculiar drama of the Pteridophytes to produce the immense wet-footed forests in which nothing could completely decay. The vegetation which died in these huge soggy tracts millions of years ago merely half-decayed: much of the sun's energy which they had absorbed and turned into carbon by photosynthesis remains locked up. Peat-bogs (see Chapter 7) do precisely the same. A peatbog is composed of a mass of small mosses whose death does not cause total decay, but rather a kind of compacted preservation. In a coal swamp, there were whole immense tree trunks, some of them 130 feet tall. Since the immense coal-forming forests of horsetails, clubmosses, and ferns, there have developed far more subtle and richly exploitative life-forms. Plants and animals have come a long way since the Pteridophytes held sway in the world.

Whenever we are near the water's edge or in soggy places we are reminded of the antiquity and vitality of the creation which surrounds us. Acid bogs, for instance, are places whose development began in the immediate post-glaciation period, perhaps 10,000 years ago, and which have continued their life-form since then with pretty well no new tricks. Formed by plants whose origins are very ancient, they are a living testimony to life forms 'designed' perhaps 600 million years ago and still doing well today. The acid bogs of Scotland are not relics of some defunct life-form preserved by an act of random good fortune. They are not museums: they are simply the best exploiters of a mean environment which happens to be sodden.

In other wetland places, there are other and sometimes complicated stories to be told. On many there will certainly be a higher preponderance of 'modern', advanced plants – the vasculars – which are probably younger than 500 million years old. Even so, we are always dealing with botanical history. In a little soggy place in the middle of Birmingham, you can find the wood horsetail (see Chapter 6) which is a non-flowering vascular plant of great antiquity. It is a symbol of the continuity which the wetland habitat represents. Bronze Age man knew the small wet wood in what is now the second city in the country. Housing development very nearly swept this wild, ancient place away, but a local campaign saved it.

There are few horsetails at the Smigel, but it remains an extraordinary record of most of the antiques of the plant world. It has, more than most wet places, a combination of what were perhaps among the first colonizers of land: the lichens, liverworts, and mosses. And, as we have seen, it has some of the ferns and clubmosses which – in their giant forms – became the first forests. On a mineral-rich trickle of water weeping its way out of a rock fissure, an algal flow grows hand-thick – proof that ancient life-forms are not extinguished by later evolutionary developments.

The lichens growing on sun-warmed and spray-drenched rocks testify to the colonizing power of 'lower' plant-forms. In many environments it was lichens which first gave plant-life a foothold on rocks, and which therefore were the first soil producers on terribly barren ground. There is fir-clubmoss, forming miniature versions of the great primordial forests which gave us coal. In every sort of damp crevasse, there are liverworts, including *Riccardia pinguis*, another of the 'primitive' plants. There are dense, moleskin-smooth clumps of the moss, *Amphidium mougeoti*; another, *Fontinalis antipyretica* – willow-moss (be glad of it: not many mosses have English common names) – clings on to boulders and can grow wholly under water

or stand periodic drying-out. Many other plants grow on the streamside: bell-heather, ferns, grasses, crowberry – even the occasional rowan, or mountain ash, clinging on to the higher sides of the ravine. The presence of bell-heather and carnation sedge indicates what is hardly surprising, that this is a damp area of thin soil which drains very quickly. With sides which are nearly vertical, this valley must be among the quickest-draining sites on earth.

After a while, the depth of the water and the steepness of the stream's rocky banks forced us, scrambling by way of handfuls of heather, up the sides. We were soon tempted to scramble down again. Should you have the good fortune to follow the Smigel inland a couple of miles, you will see that its rocky course suddenly becomes dramatically con-

torted. The waterfalls are spectacular: not vast, but astonishingly sculpted, as though a fairground designer had teamed up with a playful geological god. Deep in a pocket-sized gorge is a series of deeply cut ferocious waterfalls, frightening, noisy, dashing places where, even so, lichens spread a soft, bright presence over the rocks.

The Smigel is one of thousands of uncelebrated British streams where man's hand is hardly in evidence at all. It shows us that wetlands are living proof of the antiquity of plant-life. It shows the diversity of life-forms which have found a way of living in apparently very inhospitable wet environments. This book is about some of these places. It aims to show that they are beautiful, valuable, and often highly productive.

The River Bure, part of Norfolk's Broadland world, at dawn in September. Until the last couple of centuries man's activities here were well matched with the needs of wildlife: now the balance must be fought for by conservationists.

CHAPTER TWO

Norfolk's Broadland

The Norfolk Broads are among the sweetest, and are certainly the soggiest places in Britain. They are quintessentially the place where a low terrain and huge quantities of water are blended together. For generations people have trekked there in search of peace, quiet, and undramatic messing about in boats, surrounded by the kind of countryside where every month is a wellington boot month. They are a 150-mile patchwork of rivers and fifty lakes (broads) where ordinary civilization is left behind. They are Britain's own and only Florida Everglades.

In the characteristic broadland picture, gentle, wide rivers flow between reed banks and open up suddenly on to broads, as a grand hotel corridor opens on to vast public rooms. The broads are shallow, and often fringed by trees. They seem enclosed, secret places. Though about a million people visit them every year to desport themselves on the water – a quarter of the holiday-makers staying a week or more – almost everyone, however familiar he may be with the place, feels as though he is exploring them for the first time. Perhaps this is because you can only get to many of the broads by boat. The broads are surrounded by land so drenched and peaty that no one has considered putting a road down to many of them, and others have only cinder or wood-chip tracks going anywhere near them.

Take a canoe out at dawn from Horning, on the River Bure eight miles north east of Norwich, and there is a special tranquillity. You paddle along beside little cricket-pavilion-style waterside holiday homes with names like 'Herondelle', suggesting lovely wine-soaked evenings with the grey, tall bird, thin as an American senator's wife, etched on the gloaming. In summer your only company will be the darting, exploring moorhen, the coot, or the great crested grebe, all of which thrive beside reed-fringed rivers and lakes. Perhaps you will meet an exceptionally early-rising motor-boat hirer with a line dangling over the side. His muted "Good morning" will be in any accent of the nation; Bolton, Glasgow, or Hackney will come echoing across the grey of the water's stillness. Typical of people who take holidays on the brink of eccentricity (and the Norfolk Broads do, just, still qualify), he will be amiable to the point where you forget that you were in search of peace and quiet.

At that time in the morning, herons seem to stand guard hour after hour, only taking to their long, loping wing-strides if you approach. As for the coots, they – in the September dawns when I last saw them – are far too preoccupied with teaching their young how to sub-aqua without gear to bother with a canoeist or two. The young are done up in a lovely grey-beige and they watch their parents, who manage three-inch silver fishes in their beaks, and know the kind of envy that the early morning hire-cruiser fisherman must also feel. Fishermen say that the fish on the broads are either very bright or very few. Only the eels seem abundant, caught by professionals in battered boats going out at night and returning in the morning to collect their netted harvest. Big Dutch eel road-tankers grind along to Horning to take some of the catch to the Netherlands. Much of the remainder finishes up in east London eel and pie shops. But the coots seem to do well enough. Little families of them idle about in the rising light, darting out in convoy from a reedbed bastion or, more bravely, allowing the clan to be split by the canoe's passage with hardly any demur. Suddenly one will take the plunge and you are left craning around (not a wise manoeuvre in a canoe) for the spot where it will break surface again, with or without a fish to show for it.

To paddle into Little Blackhorse Broad, more properly known as Little Hoveton Broad, you pass through big double farm gates, swung back to let boats through; they are like gates on land, but here are designed to keep boats out in the winter. Before the Second World War, several owners of broads tried to keep the public out altogether, but the navigation interests, with their own board and constitution and the authority to make by-laws, kept them at bay. Now, certainly in the summer months, you can come and go as you please in many of the broads. You make your way down a small aquatic corridor, with an everglade effect all around. Alder trees (which, like willows, are quite happy to

The rivers and broads of North East Norfolk. 150 miles of rivers and 50 shallow lakes: visited by a million holiday-makers a year, a quarter of them staying for a week or more. Even so, much natural beauty survives.

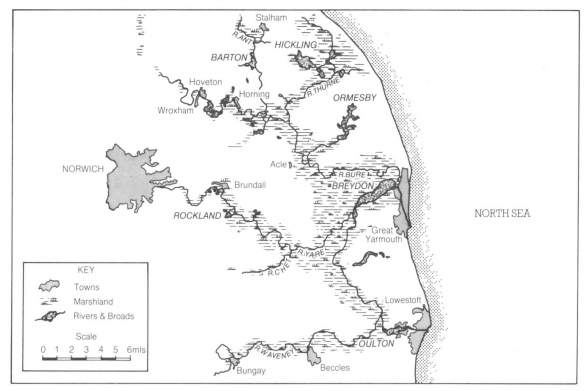

grow with their roots totally immersed in water-logged mud and peat) and a mass of grasses, sedges, mosses, and ferns among the puddles in the peat. The fringes of the rivers and corridors of the broads are barely penetrable. Even if you could get into the sticky interior, among the midges and mosquitoes and ferns, you would be at risk of sinking up to your knees or further. The broads are protected on their landward sides by a very inhospitable environment indeed.

In the broad itself, very few water-plants grow in the shallow, murky water. There is a small fringe of reed-swamp (though it is sadly

missing now from many places), and a thin band of fen beyond that, with a scattering of yellow loosestrife and ragged robin brightening the green. Yet down the river a stretch, there is a tantalizing glimpse of another sort of broad altogether. It is a private broad, but nosing the canoe past the overgrown entrance,

one can see lush water-lilies growing out over the water. The place is a wet garden. There are a handful of others like it in broadland, but they are few and far between, for things have changed radically on the broads during this century, and perhaps especially since the war. Like so much in the British natural and wilderness landscapes, the present broadland patchwork of broads, carr woodlands, and their surrounding fens are far more the result of the work of man than might be supposed.

As the glaciers retreated from Britain around 12,000 years ago, they left, in the eastern region, immense areas of flatness. There were very much bigger than they are now, extending then to what is now the Netherlands, since at that time Britain was joined to the Continent. The area now called the Fens in Cambridgeshire and Lincolnshire was almost uniformly flat, while in Norfolk, the area which is now broadland was and still is gently undulating. By the time Britain had separated itself from the mainland, the growth of reed-swamps in the soggiest areas alongside the meandering rivers in the region was much the same as for the Somerset Levels, and similarly interrupted by periods of marine transgression.

Brundall. One of the dozens of boatyards which are home base to the 5,000 motorboats which are just one piece of the jigsaw of pressures which are putting wildlife at risk. This is a very delicate patchwork of wateriness: in its way a kind of green Venice.

How much acid bog formed in the broadland region is not clear: probably not much. That man colonized this inhospitable region and saw its potential for food production is clear: such places were too rich in wildlife to be ignored. Some time more than 1,000 years ago, man had probably already cleared much of the wood – probably mostly alder, with some oak in very well-established, slightly drier places – for fuel and shelter-building. He might have been grazing his animals on drier places, and begun to manage the great reed-swamps for the reeds and the drier fens for saw-sedge and marsh hay.

Medieval man began to cut the peat which had accumulated. He may have gone for the few areas of acid, sphagnum-bog peat first, since this burns best, and then moved on to the poorer, shallower, fen peat afterwards, especially when the deeper workings became too susceptible to flooding. Certainly, fen peat was being cut up till just after the First World War. In digging out two-, three- and even four-metre 'open-cast' peat workings, he was producing the final part of the broadland landscape. He was adding the shallow 'lakes', or broads, to the existing scene of river, alder, fen, and marsh. In the fourteenth century, a raising in the level of the North Sea added very considerable flooding to the water system. That, and the cutting of dykes between various of the broads and the rivers, completed the development until, centuries later, in spite of an increasing problem of flooding, it became economic to drain the marshes which lay in the south and east of the region. Building tidal embankments to the east and running dykes through soggy, poor fenland produced grazing marshes which flooded every winter when the rivers could not drain all the fresh water away, or when the sea made incursions. The river flooding brought refreshing doses of silt to the marshes.

It was early-nineteenth-century technology which accelerated the changing look of the grazing marshes – which run like webbing within the fingers of the Yare, Bure, and Waveney rivers – with the new potential to lower the water-table which wind pumps brought. Our own century took the process a stage further with reliable steam and diesel engines and electric motors, so that in some places the ancient grazing marshes could be dried out enough to produce cereal crops for feeding cattle indoors. The change of use seems ordinary enough, but it takes a wet environment, with ditches full of flowers, and shallow bird-feeding 'lakes', and turns it into ploughed arable 'prairie'. (See Chapter 3.)

The changing uses of broadland

The medieval peat-workings had been dug to depths of up to five metres by brave or desperate men, since there was then – and the theme continued well into the nineteenth century – nowhere quite so diseased and perilous as these dank, insect-ridden regions. Norfolk reported the last case of leprosy in Britain and was the source of a home-grown malaria. Norfolk, as recently as a century ago, was as tricky a place as any far-afield eastern outpost of the Empire depicted by Graham Greene or Somerset Maugham.

Somewhere around the fourteenth century the sea level, relative to that of the land, began to rise. The workings flooded to create small lakes. On the land around them, which must have been a paddy-horror of dampness for generations of people who knew nothing of rubberized waterproofing, or of immunization, a peasantry devoted itself to making

---

The dawn loveliness of Little Black Horse Broad (properly called Little Hoveton Broad). Edged partly with carr wet woodland on peat, and fringed with reed swamp and sedge fen. But this waving stand of reed is a fraction of its former self.

some kind of arthritic and rheumatoid living from the waterlogged soil. There was enough work for any number of people, but probably precious little profit to be had. To the great landowners, profit perhaps meant rather little; it was the lowlier peasant who bore the brunt of the work and made a pathetic living from his toil. There can hardly be anything less healthy in the world than working among the harsh east winds which Nelson, born in Norfolk, so detested, and which made him love the tropics. Rheumatism must have bent more knuckles in our wetland countryside than anywhere else in the world.

There was the rich marsh hay to cut in the summer, for fodder. Wherever a spot was left alone, unless it was ruinously soggy, carr would eventually invade the fen and provide the usual crop of small poles and timber that any wooded area provides. Here, though, it would be the kind of tough wood which makes ideal broom-handles, and stakes for

uses where impermeability to water is an advantage. Alder stakes are now beginning to be used as riverside piles in place of the ugly and expensive steel piles or imported timber which were in widespread use until recently. The fens grew two distinct sorts of thatching material. The reed for which Norfolk is famous was encouraged to grow in huge areas. It is pricier but it is also twice as durable as straw for roofing. There was also sedge which was and is used for capping thatched roofs. Digging for peat continued, though there was mostly only poor fen peat to take; the workings were shallower than ever, and their produce less and less good for burning. They flooded immediately, and were colonized by the common reed, *Phragmites communis*, and other reed-swamp plants. Pretty soon, many of these shallow little 'lakes', or turf ponds, disappeared, grown over and filled with dead vegetation until it was impossible to tell whether they were lake or swamp, while at their fringes, fen, dominated by sedge, would invade. The broads, more than any other part of our landscape, are in constant flux.

It required great industriousness of many workers to keep nature at bay in the broads. For centuries their task was achieved by heavy cropping in the areas which served as fields, and by pruning and coppicing in the woodland areas for small poles and the split saplings which made the wattle part of the wattle-and-daub houses of the centuries before brick was in common use. Plenty of

other plant-life was just as greedy to live there and compete with a farmer's plan. Alder jungle was always threatening his fenland economy. There being no such thing as a natural state for the broads, every generation decides what is ideal, workable, or possible. For us that means a state in which a large number of species (especially rare and fragile plants and animals) is sustained. Only by such indicators can we feel that we are maintaining at least a part of our world as a bastion for the kinds of life-forms which need an environment unsullied by man's predations – or at least one where we manage things to the best advantage of non-human forms of life.

The broads that people began to sentimentalize about were those of the final years of the nineteenth century. They were already

Left. A windmill. One of the classic broadland scenes.

Right. A derelict punt which would once have been invaluable to a fen-worker, getting around along the rivers and dykes of the broadland wet farms, and bringing out a harvest of reed, sedge, marsh hay, and coppiced wood.

damaged by the time of the Second World War. In the first decades of this century there were still considerable numbers of agricultural workers on the land. They might spend their nights eel-fishing (as several people, one or two of them very crusty characters, still do) and their days fen-farming on their own account, or they might be fully employed by one of the great farming estates, such as the estate which owned Woodbastwick Fen, now incorporated as part of the Bure Marshes National Nature Reserve managed by the Nature Conservancy Council. In those days, Woodbastwick Fen was providing several different crops, and about sixty-four of its 232 hectares were open water. Now, only continuous work by the NCC staff and volunteers is rolling back the intrusion of small

but powerfully sturdy and invasive alder trees
and the like. The fen soggy-scape would easily
and rapidly be overtaken by exuberant damp-
loving plants and trees. Eventually – surprisingly quickly – there would be no open
water left, just scraggy woodland. In the
meantime, the NCC is looking forward to its
commercial crops of thatching materials.

The farming carried on in Norfolk's
fenlands was unlike any other. It might depend
on marsh grass, or sedge, or reed, or even on
letting everything go until young trees invaded, and then cropping those. Wildfowling
was a profitable sideline for estates, especially
in the years when the British public ate more
wild game than poultry, and the battery bird
had not been invented. There was no need
for planting, ploughing, manuring, or even
conventional weeding, only the endless business of keeping out the invasion of unwanted
scrub.

Dr Martin George, the Nature Conservancy Council's regional officer and an acknowledged expert on the ecology of the
Norfolk Broads, has been studying some of
the historical evidence about broadland use.
"Things changed very dramatically because
of the First World War," he stresses. "So many
of the young men from here went away to
fight and did not return. After the war, there
really was not the kind of labour force that
there had been before."

The Edwardian period was also the last
period of the great landowner. At that time,
many of the great Norfolk and broadland
estates were owned by shooting and fishing
people who combined a passionate interest
in wildlife with a powerful urge to kill it.
They had no inkling that many of the birds
they enjoyed shooting would shortly be on
'missing, presumed extinct' lists. Indeed, in
the first years of the century, some rare water-
birds were making something of a comeback.
But the great estates were running out of
money, and parts of them are now more often in the hands of the Norfolk Naturalists'
Trust than those of aristocrats and magnates.
Plutocrats and peasants are in short supply
in modern Norfolk, but there are tourists,
farmers, water authorities, and naturalists in
plenty, and they all want different things of
the land and water.

The Bure River in the early morning. Long before
the 'rush hour'.

The rivers and broads, once clear and rich in varied life from top to bottom, are now only occasionally as interesting as they were in the 1930s and hardly anywhere as fine as they were at the turn of the century. The marshes are almost everywhere much drier than they were before the Second World War, and the fenland is only in good heart where conservationists are making a conscious effort to keep it so, while there is little or no coppicing of the alder carr as there once was. One reason why the plant-life of the broads is much poorer than it once was becomes apparent every summer morning, especially on the busiest river of the three, the Bure.

You have been out in your canoe, with the sound of water dripping off your paddle and licking and slapping at your bow as it cleaves the still broad. The place belongs to you, to the herons and coots and moorhens, and to the early-rising fisherman. Suddenly, as though to a word of command, at nine o'clock, there is an instant rush hour. It does not last for only an hour, but goes on till late afternoon. There are more than 2,000 motor boats for hire on the Norfolk Broads, and around 3,000 in private hands, and many of them spend their days surging up and down the rivers and broads. There is a speed limit of seven miles per hour (five mph in some places), but it is often ignored. The wash from boats slaps and bangs against the fragile mud and peat banks of the rivers, wrecking

Below. A heron, relaxed and hunting in the reed-bed.

Right. A broadland converted windmill. It was the coming of the more powerful steam and diesel, and now electric, machines – and enclosure – which made the drainage of the surrounding marshes possible.

the reed beds with their flowering rush and bur-reed which might alone defend them, and stirring up the sediment in the shallow water. Prime victims of this process are the white and yellow water-lily, and arrowhead with its white flowers. Aerial pictures suggest that before the war there were around 1,200 hectares of reed-swamp in the broads. Now there may be as little as 600 hectares.

Nor is it the big, live-in boats alone which cause the damage: there are small boats, hired by the day or the hour and driven, Dr George suspects, by people who understandably want to get the best out of the short spell on the water. "The smaller boats go quite fast," he says, "but what's worse, they make a disproportionate amount of wash." He hopes that one day the speed limit will be lowered and more strictly enforced, and that the governors on boat engines will be made fiddle-proof. The odd thing is that there is only one place on the broads where boats need to go at all fast. This is at the junction of the Bure and the Yare at Yarmouth, where the current is strong for about 200 yards at certain stages in the tide. Boats could be asked to wait till the tide is slack, and then need never go faster than four mph as they do on the canals.

Boats are ruining the banks of the rivers and threatening the reed-banks. They are stirring up sediment which moves with the rivers' currents into the already shallow broads, where the silt settles out, making the broad too shallow to provide a good environment for many interesting plants. The motor boats' propellers chop up the young shoots of plants; the shallower the broads, the nearer the plants on the broad-bed are to the swirling blades. The quality of the river banks, the sediment making the broads shallower than is good for wildlife, the constant disturbance by propellers – all are good arguments for restricting boat hire and the type and speed of craft that should be allowed. They all argue for a return to the sailing craft which wiser and quieter spirits still keep or hire on the Norfolk Broads, and which predominated until after the Second World War.

The boats are merely the most obviously intrusive danger. The broads are also the sump into which are poured much poorly treated sewage and leached-out fertilizer from nearby farmland. The plight of the broads springs in part from the historic use of the rivers as cheap, open-air sewers, in part from the problems that modern farming methods impose on our waterways.

Sailing boats on Hickling Broad, one of the broads which conservationists have managed to influence for the better.

Understanding the broads

Modern ecologists have charted a system for describing deterioriation in broads waters. It effectively describes the stages in which these chalky, calcium-rich waters are affected by successively larger doses of the two great growth-promoters in nature, phosphorus and nitrogen. Both of these have, through man's agency, been introduced in greater and greater concentrations into the broads. The increased doses of each have radically altered and recently, in many places, wrecked the wildlife of the entire region. They have proved an embarrassment of riches, resulting in nutrient enrichment, or eutrophication. Without them, there is no life. With too much of them, there can be almost too much, and mostly of the wrong kind.

Overleaf. Ormesby Broad

A sailing boat on the River Ant. If more people could be persuaded to holiday under sail, as the original holiday fans of the broads used to do, the aquatic ecology of one of our loveliest regions would be richer.

After the Enclosure Acts of the early nineteenth century, and with the new technology introduced then, it became possible, legally and technically, to install more wind pumps to drain the marshes in the region. Recent research suggests that slightly more phosphorus and many more nitrogen compounds must have found their way into the broads during that period. This is just what abundant plant-life would have needed: one part of phosophorus to ten parts of nitrogen is a most potent fertilizer for plant growth. The years between 1800 and the beginning of the First World War might be identified as the richest period for the broads. There remained a peasantry to work large areas as wet agriculture, and enough wilderness to provide an astonishing variety of bird and insect life. Some of the broads themselves were like perfectly fertilized water-gardens, sporting exotic growths of lilies. In many of the broads there would still have been an abundant 'bottom' life, with Canadian pondweed, hornwort, and perfoliate pondweed all giving food and shelter to small insects and animals which were a useful part of the diet of fishes, and so of birds.

By the 'thirties, when the sewage works of an increasingly populous and urban Norfolk were beginning to pump all too much phosphorus into the water system, there began to be a real anxiety that the exuberance of the growth would choke the broads altogether. The shallow lakes were thick with white and yellow water-lilies. Below the surface, things were not thriving so well, often shaded out by the proliferation of the plants above. The broads were a fisherman's and bird-watcher's (or wildfowler's) paradise, though the yachtsmen were beginning to complain that they could not always sail as freely as they would like.

Pictures of Alderfen Broad from 1930 and 1975 tell the story. The former show dense lily cover from edge to edge, making the lake look like a well-stocked garden pond.

By 1975 there is open water. At some other places the decline was even more dramatic: even in 1960, the River Commissioners were removing from Hickling Broad 2,000 tonnes a year of water-plants which would otherwise have choked it entirely. Again by 1975 there was nothing to impede the boaters' way, though in the case of Hickling, the exploding post-war gull population, part of which roosted at and excreted into Hickling, was taking its toll. There and elsewhere, algal blooms, exuberant growths of phytoplankton, were thriving in the enriched waters, and were taking all the available light. They were also impervious to the thrashing boat propellers.

But if there was no over-abundant water plant-life to worry about, there was a dramatic increase in the amount of sediment which was being brought into the broads. Many were becoming shallower by several millimetres a year, rather than at the far slower rates of a hundred years ago. Ordinary silt is an element in the material accreting on a broad's floor, but the dead bodies of algae and the calcarious marl precipitated out of the alkaline water by the algae can often add up to well over half the total.

In some of the lodes (dykes) of the Bure Marshes, conservationists are discovering more of the effects of cleaning up the water: here water soldier and lily pads are thriving in the clear, undisturbed water.

By the Second World War, and at an ever greater pace since, the broads were falling out of phase. Boats, and later – after escapes from fur-farms in the 'thirties – grazing by coypu were knocking back the reeds which gave shelter to many small animals and which held the banks of the rivers and broads in place. Many birds, driven to unhabitual grazing habits by shortages elsewhere were also devastating the reed-beds. By the 'sixties, the only species which did well on most of the broads were those with remarkable staying-power, or those which had the weed-like capacity to live opportunistically. The moorhen, coot, heron, and grebe managed to hang on nearly everywhere.

The decline afflicts almost every species. In some cases creatures and plants that might otherwise be common become rare; in others, what was always rare becomes non-existent. Dr George wrote in the mid-'seventies: 'Water-bed weeds have disappeared altogether from most of the broads... The invertebrate fauna dependent on water weeds has been greatly impoverished... Mean numbers of tufted duck wintering in Hickling Broad declined from 160 between 1950 and 1956 to 31 between 1976 and 1979.' The names on the 'very scarce' list make a beautiful and a sad litany: fen orchid, swallowtail butterfly, marsh harrier, bittern, or broadland's unique dragon-fly, *Aeshna isosceles*, a hawker dragon-fly now confined to a few dykes of the disappearing unreclaimed grazing-marsh dykes and one fen site.

As the marshlands suffer increasing drainage they become less useful as feeding-grounds for marsh harriers and waders, and less habitable for wintering wildfowl. As the reed-swamp becomes scarcer and more open, the shy bittern finds sanctuary harder to find (this also affects the marsh harrier, which nests in reed-swamp). Black tern, which likes to nest in swampy ground, only visits the broads on migration, while the white-winged black tern is even rarer. Even the mallard do

less well than they used. Sparrow-hawk, however, are beginning to recover as a result of protection by legislation in the nineteenth century, and are now becoming commoner in the broadland carr. The marshland dykes

which should support dragon-flies, damsel-flies and the spiky-leaved, white-flowering water soldier are all more or less threatened with pollution from the fertilizers and pesticides which modern farming likes to use.

A kingfisher, an almost heraldic symbol of British wetlands.

Restoring the broads

As you cruise along the broads there are tantalizing moments when one of the river-corridors opens into an alley of water whose passage is barred after a few yards. Sometimes it is a tiny private broad with barred entrance, and sometimes the owner can rely for privacy on the impenetrable barrier of carr. Recently conservationists, co-ordinated by the Broads Authority and its new 1982 Strategy and Management Plan for Broadland have taken a hand. At Cockshoot, they have dammed-off a pretty little broad just downstream of Horning. The Broads Authority and its conservation officer, David Brewster, are trying to discover whether keeping a broad free of phosphate and pumping out the mud and silt so that the water depth increases and brings about a real increase in the plant and animal life.

The experiment is crucial. The water authorities who handle and have jurisdiction over water both before it goes into people's taps and after it comes out of their sinks and lavatories have already agreed to costly phosphate stripping on the River Ant, but Barton Broad, which is fed by the Ant, is not yet showing enough improvement to make a cast-iron case for phosphate stripping elsewhere. To be rid of some of the nitrates would be theoretically useful but virtually impossible. David Brewster points out that the broads receive around two thirds of the water falling on Norfolk. That water consists of rain falling the length of the county on fields replete with modern fertilizers. The catchment area of the broads is something like 3,600 square kilometres. Unfortunately, no one has yet developed fertilizers which stay where they are put. The tonnes of nitrates farmers spread on their fields seep straight into the nearest water system.

The race is on to learn how to save the broads. At Barton Broad, the second-largest in the system, with 180 acres, the Norfolk Naturalists' Trust and other conservationists have persuaded the Anglian Water Authority to keep the River Ant rather cleaner of phosphates for the time being, since they believe that phosphates are the crucial key to modern

Common reed. Lovely still, but its acreage is down by half since the Second World War.

eutrophication. For the summer growing-season of 1982, conservationists transplanted 16,000 water-weeds (mainly water soldier and water-lilies) into the broad in the hope that they would mop up previous seasons' phosphate-residues which were still being released in the mud. Progress is slow. In 1982 the phosphates were still too rich, even though the main river influence on the broad was so much cleaner. The belief now is that in successive seasons the phosphate-store in the mud of the broad will become sufficiently diminished to give macrophytes a chance. Nonetheless, the conservationists are in effect on trial. They must come up with results to give the policy-makers confidence that they at least have the techniques to bring life back to the broads, if they are given the resources.

The conservationists are anxious that the Anglian Water Authority should keep its enthusiasm for the phosphate-stripping operation, because it is something within our technical and financial grasp. Besides, seventy-five per cent of the phosphates in the broads come from easily regulated sewage works. The water authorities have already agreed to costly phosphate-stripping on the River Ant, and decisions need to be taken soon about other rivers. Phosphate-stripping does seem, so far, the most accessible policy.

Naturally enough, David Brewster is very concerned with what goes on at Cockshoot. The assumption here is that mud-pumping will remove from the broad most of the phosphate-rich layers of mud. Since the dam ensures there is no eutrophic water coming in from the River Bure, it should be that within a season or so the lilies will come back. Their reappearance would go a long way to prove the phosphate connection. At the eight-acre Alderfen Broad, where phosphate-rich waters have been diverted to bypass the broad but where no mud was pumped, recovery began to take place after about four years. This suggests that the phosphate-stores in the muddy depths of a broad can last that long, though their persistence depends on many factors, such as how long the broad has suffered eutrophication, and how frequent is its water 'turnover'.

"We dammed Cockshoot Broad off last year, and put in a board walk along the dyke which joins it to the River Bure," said David Brewster. "No one had been in here for years, it was so strangled with a tangle of carr. It's obviously not something we can do on every broad, because we need to block it off from the river, but it will give valuable insights into the recovery of broads. We pumped out 40,000 cubic metres of mud, and that gave a water depth of about three feet instead of the six inches or less there was before." Like plant archaeologists, the conservationists were also pumping away younger layers of mud, and getting down to mud that was twenty or more years old. As expected, there were still-fertile seeds of many interesting and richly various plants. "We found all sorts of things, especially in the dyke," says Brewster. "I've drifted along here in the boat with a face mask on: you can stare down and see the shoots of yellow water-lily and the more uncommon white water-lily, and there was Canadian pondweed and bladder-wort too."

Bladder-wort is a tall-stemmed carnivorous plant with great free-floating submerged roots which look like a moss and which trap tiny water-animals to feed the plant. Small yellow flowers appear in July and August. It is rather rare now, and a good indicator of things going well in a water system. It hates water which is too rich in nutrients. Bladder-wort is the perfect example of a plant which thrives in a complex environment which is relatively undisturbed. It is indicative of other high forms of plant and animal life. It hates being churned up by propellers, is easily starved out of existence by greedy algae which steal the light it needs, and hates shallow dirty water. Its appearance on Cockshoot is a good omen.

The process of improvement is not a simple one. The River Bure may no longer be depositing its load of silt and phosphates and nitrates into Cockshoot but there are still plenty of phosphates in the silt, and they may take several years to be used up by plant and animal life. In the first year of the experiment, it seemed that zooplankton began to gobble up much of the phytoplankton (algae) and that the algae would be eliminated. But by late summer the phytoplankton, warmed in the summer days, regained a temporary grip on the broad. The shoots of higher plants were still there and have continued to survive, but it was clear that phytoplankton were sturdy creatures and would quite probably find the high doses of phosphates they need for a couple more seasons as it is released from the silt. Only when the phosphate levels are low enough to starve out the phytoplankton will the water-plants be able to play their trump card.

Water-plants are not only good at remaining fertile though dormant for decades; they can store energy for a season or more. With luck there will come a time, this year or next, when the phosphate levels are very low and phytoplankton cannot survive. Then the water-plants will come into their own, using energy stored from the year before. Once the water-plants have vigorous life, it

will be the phytoplankton that cannot compete. When that happens, where there are
abundant higher forms of plant-life there will
also be a rich habitat for small insects, worms,
and snails. Who knows, it may even be that
fresh-water mussels (there are a few empty
shells of these on the floor of the dyke at
Cockshoot) could be re-introduced.

Then it may be that white and yellow
water-lilies, great bladder-worts, other flowers
and dozens more bright, rich plants will be
in evidence, as might a richly revived variety
of insects and birds, the entire scene fringed
by great waving armies of reed-bank at the
edges.

Common reed, in early autumn, when it is in
seed.

Overleaf. The mouth of Cockshoot Broad.
Entrance is barred because, though
conservationists have opened up the broad, for
the time being they are keeping boats out whilst
its health is restored.

The glories of broadland: places to visit
The Norfolk Broads are a threatened environment, with none of them being free from damaging eutrophication and deterioration. Yet there are several places where reserves of various kinds have been set up, and many of them provide splendid opportunities for the public to see wildlife and the older habitats of broadland.

One of the most famous of the broads, Hickling, makes a fascinating case history of how interest in the conservation of nature altered during this century. Three liberal politicians, Edwin Montagu, Sir Edward Grey,

and Lord Lucas, took over White Slea Lodge Estate and one or two houses in the neighbourhood of Hickling Broad. They were the traditional, nature-loving, shooting sort of gentlemen: but – and this seems to have been specially true of Edwin Montagu – they came to realize that their love of birds would be better expressed by conserving at least the rarities among them. They went on shooting coot and various duck, and gave them to the villagers to eat. They foreshadowed more recent legislation by refusing to shoot many predator birds and other scarce species. Indeed, they employed a noted local youngster, Jim Vincent, who was to become a famous ornithologist. He was part gamekeeper, part conservator, as well as being something of an author. Perhaps at Edwin Montagu's instigation, Jim Vincent kept a diary which was illustrated by George Lodge and has since been published (*A Season of Birds*, Weidenfeld and Nicholson,).

In his lifetime, Jim Vincent recorded 247 species of birds at Hickling. Some of those had been attracted back to Hickling after many years of absence, probably as a result of over-shooting. In 1910, the men found the nest of a Montagu's harrier (the bird had been named after a nineteenth-century Montagu, not Hickling's). The nest contained an egg which hatched what was perhaps the first-ever breeding Montagu's harrier in Norfolk, and began a tradition which has not stopped since. The ornithologists mounted guard over the nest and young. In 1915 marsh harriers returned to breed, instead of merely being visitors. The bearded tit, sometimes called reed pheasant, also established itself more strongly in these years. In 1886 the last bittern to be seen in Norfolk had been shot. The first reappearance appears to have been noted by Jim Vincent and a nature photographer, Emma Turner, in 1911, since when the bittern has become one of the prized broadland birds.

The Yare River, showing its continuous fringe of carr woodland, a wet-footed world which does a great deal to make the broads rivers feel quite separate from the ordinary world of agriculture around.

The syndicate which had saved Hickling and ensured that its broad and surrounding were managed for conservation were lucky to have their work carried on by Lord Desborough. In 1926 he bought the estate, following the untimely death of his son, the Hon Ivor Grenfell, who had taken a share in the property's rent a couple of years before. In 1944 the Desborough family sold the estate to the Norfolk Naturalists' Trust, the first of its kind, formed in 1926. By 1958, 1,204 acres, including around 370 acres of open water (Hickling is the biggest of Norfolk's broads), was designated a National Nature Reserve.

Hickling Broad has for many years been relatively fortunate in conservation terms. Besides being well looked after, it has the advantage of being almost at the top end of the river and dyke system which feeds it, and therefore does not receive quite so much silt and enrichment as some others. It has its difficulties nonetheless, and is noted by the

Dusk on the River Ant at How Hill.

Broads Authority as being in a condition which is transitional between mild and extreme enrichment: in other words, even the coarse aquatic plants may be under threat. One of its difficulties is unique: more than its fair share of black-headed gulls roost on it, and their excreta are a potent source of enrichment.

The water in Hickling is slightly brackish because of seepage from the North Sea, which is only three miles away. Indeed, the land around Hickling has always been subject to North Sea flooding, and this leads to a chemical problem which is affecting the broad. Although most of the surrounding marshes have been drained with varying degrees of thoroughness by diesel or electric pumps, there are specially engineered lagoons which imitate the pools once left by relatively inefficient wind pumping. Heron, lapwing, redshank, snipe, ringed plover, teal, shoveler, gadwall, and mallard can be seen at most times of year from Deary's Hide, on the marshes to the north east of the broad. Spotted redshank, greenshank, sandpipers, dunlin, ruff, and yellow wagtail also visit seasonally.

The reed-beds at Hickling are only partially cut, in late winter and early spring, for their commercial crop used for thatching. The small, tawny, bearded tit (the male has a striking lavender-blue head with the eponymous large black moustache) and the reed bunting (the male has a black head and white collar) make their homes in the areas left uncut. In the drier sedge-beds, which would be very like the antique fens if it were not for the sedge harvest every three summers, there are large numbers of dragon-flies, damsel-flies, and butterflies and moths, including the beautifully named small marsh footman.

The fen plant community boasts some very pretty plants, but among them is the humble-sounding, rather ordinary-looking though actually rare, milk-parsley (it is only found in eastern England and in Somerset), the main food-plant of the swallowtail butterfly, of which the only British colonies are in broadland, and which is a prime symbol of the wildlife significance of the region. The swallowtail uses the milk-parsley as a resting place for its eggs, the caterpillar feeds on its leaves, and the adults feed on its nectar. There are many of the fen plants in these sedge beds: hemp agrimony, yellow and purple loosestrife, marsh bedstraw, great water dock, and lesser reed-mace (otherwise, though wrongly, known as bulrush).

On the water of Hickling Broad itself there are the shallow-water species, whose feeding method is to dabble at food by bobbing down into the water while staying afloat on top of it: widgeon, mallard, teal, gadwall, and shoveler. Among the divers to be seen feeding in the deeper parts of the broad there are pochard, tufted duck, goldeneye, and scaup, with pochard and tufted duck often staying the whole year. And there are, as always, large numbers of coot.

At the Norfolk Naturalists' Trust's Broadland Conservation Centre there is a floating gallery and information centre giving views over Ranworth Broad, and an illustration of the entire succession of plant-life from the reed fringe through reed fen, swamp carr, and wet woodland, to oak woodland. In the reed fringe (there is little left at Ranworth), there are reeds, reed-mace, and rushes. This, then, provides the dryness and accumulation of peat which becomes the basis for reed fen, which is cut every year (in some places every other year) for its crop, allowing milk-parsley, marsh valerian, and great hairy willow-herb to flourish.

At Ranworth, the alder or swamp carr continues, in the classically described ecological succession, right through the stages of dryness underfoot that culminate in mature oak woodland. It has often been said that oak woodland is the climax or ultimate succession of habitat which would cover much of Britain if it were not for man's intervention. This implies, wrongly, that left to themselves many wet habitats would, by the constant accumulation of peaty die-back vegetation, become dry enough to be invaded by oak, with hazel, birch, ash, and holly growing beneath. There is evidence that many places will stay wet enough to remain as reed-swamp or alder carr for many years. A change of climate, or a rise in the sea level relative to that of the land, will alter that, of course, and in many places the accumulation of peat will create a soil dry enough to support oak woodland. The process is not inevitable everywhere, and often sphagnum moss appears to be a climax vegetation, though this appears to happen very seldom and only in small patches in broadland.

Hickling and Ranworth nature reserves can be reached by boat or by car. The Nature Conservancy Council has laid out a trail of its own which can only be reached by boat. It is on the River Bure, with a landing stage opposite Salhouse Broad, in the midst of the large Bure Marshes National Nature Reserve. The only part of the reserve which is open to the public, it is richly interesting and well worth a visit. The NCC's trail lies in alder wood beside Hoveton Great Broad, though it looks out over the broad itself and includes small beds of sedge. Its greatest excitement is the opportunity to wander through the nearest thing Britain has to tropical rain-forest. Safety considerations require you to keep to the railway sleeper boardwalk: it is an injunction few need much bidding to obey, for the swamp on either side is clearly dangerous. In several places, the boardwalk moves as you walk along it.

Alder wood will not grow where land is too acid or too poor in nutrients, which is why it does not get a hold on many of the classic bogs, even if they dry out somewhat (see Chapter 7). In the wet Atlantic period alder was probably a good deal more common than it now is. Tansley thought that the alderwoods of the broads were all but virgin woodland, having such a 'tangle of vegetation with trees of all ages, old decaying trunks remaining *in situ*, rotting logs and branches on the ground, and saplings growing up between them'. In the Hoveton trail there are places where the alders have never been coppiced: they are single-stemmed as opposed to those which have regularly been cut in the past, which throw out dozens of limbs from a single 'bole'. Wild blackcurrant grows in this stand of alders. Part of the way in which swamp alderwood sometimes dries out comes from the very success of alders in wet ground. Growing too large for their roots to support them in such soggy terrain, they fall into the sodden peat below, where their trunks and branches contribute to the soil-making process.

In parts of the nature trail there is evidence of what can happen where an 'island' of peat is left undisturbed: oak, ash, guelder rose, birch, hawthorn, redcurrant, and dog-rose all grow together in one part of Hoveton's alderwoods. Even where there has been disturbance, and the wood is relatively young, there is the expected association of alder, the shrub sallow, marsh fern, the rarer royal fern, yellow flag, and many mosses. Most of the broadland birds can be seen from the hide on the trail, and since there is some milk-parsley near the trail, one might even see the swallowtail butterfly.

Purple loosestrife, enriching the riverside fringe of reedswamp.

CHAPTER THREE

The Somerset Levels

The Somerset Levels are a crucial and beautiful part of the wetland picture. They consist of about 68,000 hectares of the flood-plains of five major Somerset rivers and their tributaries. They are much less extensive than the fenlands of East Anglia, and more intimate and welcoming in their aspect. They are more secret sorts of places, yet they are hummingly rich in life in summer, and haunting in winter.

They have never attracted the kinds of crowds that flock to the Norfolk Broads, and there is no massive tourist industry associated with them. Partly, this is because much of this countryside is not open or accessible. There are footpaths, but not the obvious cross-country routes of the moorlands and downs, nor the river or canal passages of many other beauty spots. There is no one grand, single landscape which defines the Levels, as there is in the fenlands or peatbogs. Rather, they are a scattering of separate areas associated with several different rivers running down to the Bristol Channel, but running – crucially – very slowly, and against a backdrop of steep, definite hills which provide, besides important geological barriers, viewpoints and visual longstops. The lines of hills constitute the toes and the wetland elements of the landscape, the webbing itself of the webbed foot which runs away from the Bristol Channel and the Severn estuary into the eastern heartland of Somerset.

Sweeping down on the place along the M5, which cuts its twentieth-century swathe across this ancient landscape, one is plummeted from the fast lane into a world whose pace is that of a cow's heavy-uddered, swaying plod and was once that of a professional wildfowler or a monk. A few minutes' drive from Taunton and the motorway there is a steep little ridge of hills which runs down to one of the classic Somerset Levels sites, West Sedgemoor. Even getting to it affords great pleasure: a sudden brow of a hill, and the vista is before you. The far horizon is a sun-haze away, and down below you is the geometric patchwork of fields, each carved out from the next by a rhyne (pronounced 'reen', meaning ditch). A flower-strewn, precipitous meadow makes the perfect vantage point, a scenic springtime or summer *chaise-longue* from which to view the scene. There are many such spots overlooking the Somerset Levels landscape.

Above. West Sedgemoor, from Red Hill.

Below. The controversial Somerset Levels. Properly speaking, it is the clay and alluvium areas which are called 'levels', whilst the peat areas are called 'moors'.

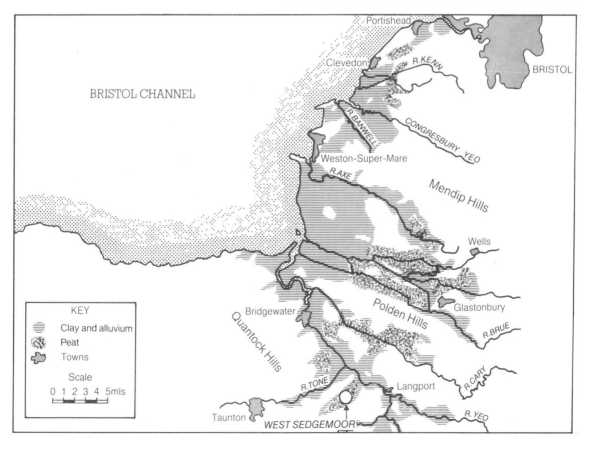

Far more than the fenlands of East Anglia, the Levels are a patchwork of higher land and low plain. When the region was wetter and wholly undrained, it provided a richly varied environment for man. Its 'islands' attracted monastic settlements whose busy superiors were keen improvers and drainers, but their early activities are now more popularly remembered as forming an integral part of the English religious mythology. Glastonbury Tor, a bleak, perfectly formed bump made of sandstones and siltstones, is so dramatic a natural feature that it was bound to attract attention. Joseph of Arimathea is said to have buried the Holy Grail there. Arthurian legends swirl round these curious higher parts of the Levels system, fostered by a literature of widely varying quality which is one of Somerset's bigger industries.

Actually, the Somerset 'Levels' is a catch-all name from the 'levels' proper, which are an estuary-bordering landscape area composed of clay. Inland, and making the floodplains of various rivers, are the 'moors', such as West Sedgemoor, and many others. Many of the moors are bounded by the small, steep hills on which farmsteads are found. The moors themselves have been used for gener-

West Sedgemoor. From a distance this looks like ordinary grazing land. Actually, of course, it is grazing land formed on ancient peat.

ations for grazing and haymaking; a farmer's holdings on the Levels are seldom the main part of his land. The moors incorporate, rather, the equivalent of an upland farmer's rights to 'walks' on a 'hill', or of an estuarine farmer's entitlement to a 'stint' on a saltmarsh grazing.

This is an agricultural environment, tucked away from much public interest or involvement in recent years, and now fiercely

contested as some – not all – of the farmers want to claim Government 'improvement' grants. The Levels are made up of a patchwork of fields divided by rhynes. There are hardly any substantial trees, but looking across the winter landscape from a distance, a fine pencil-line of pollarded willows follows the banks of the rhynes, while in summer the highly groomed branches are puffed-up green. The rhynes operate as a wet hedge, draining water from the fields in the winter and irrigating them in the summer with water which also slakes the cattle's thirst. The rhynes are dammed-up with sluice boards during the summer months, in a system of penning which turns what is drainage in winter into summer irrigation, maintaining high water-tables. The rivers of the Levels have mostly

This bull on Tealham Moor represents the livestock farming of the Levels, under threat of intensification or of conversion to arable.

been devastated in the name of efficiency. They are high-sided, wide, and functional, their banksides more populated with anglers than with trees or shrubs. Even so, they can provide dramatic walks.

The farming of the Levels depends overwhelmingly on dairy production, but there used to be a really big 'withy' (willow) business here. The Somerset Levels still provide almost all the English-grown basket-weaving materials and much of the artist's charcoal, though on an acreage which is around a third of what it was at the turn of the century. Dotted around many of the river valleys can be found the small ruins of old withy works and boiler houses, marks of a local industry in decline.

In summer, the pleasures of the Levels are deliciously mild. Perhaps this is why it has taken years for the public to realize that – quite apart from the shocking waste of public money involved (see Chapter 9) – the further drainage of these fields does matter and would be a loss. The process of drying-out the Levels had been so gradual and piecemeal, as so much environmental destruction is, that few sounded the alarm, and none

A withy crop on Sedgemoor. Once a mainstay of the district, there is much less cultivation of osiers now.

could easily make it heard. Getting acquainted with the workings of a rhyne on a humming June day soon leaves a trail of hints as to the fineness of these places. You need to find a good one, perhaps on Curry Moor, or West Sedgemoor, where there has not been much fertilizer-use. The ditch is only a few feet across and its waters are crystal clear and just a few lazily flowing feet deep. Among the likeliest plants to be seen, apart from the pondweeds, will be frogbit lily, its leaf a couple of centimetres of green, palette-shaped, with a small white flower, like a linen petticoat up-turned at the sky. More dramatic is the flowering rush, built like a long-stemmed candelabrum. A dragon-fly hammers by with its odd mixture of vulnerability and aggression, and there will often be pairs of damselflies with see-through wings edged in black, like netted mourning paper.

What looks at first glance, from a distance, to be an ordinary, straight, dull ditch, turns out to be a busy, sexual, brightly-coloured, complicated world. It is a world whose ordinariness might seem robust, but the ditch and pond environments of the British farm are among the most threatened of the semi-natural habitats. They usually contain too little that is scientifically rare enough to get into the top league of conservation requirements; but, just as the hedgerow had been all but extinguished in some regions before anyone noticed, so the unpolluted stream and ditch may soon be extinct. The child of the future may well come to relish as a rare experience the fun of messing about in a flower-fringed stream or ditch.

In winter the Levels show a very different face. There remain several moors which, unable to shed their water quickly enough to be prime agricultural land, flood. This means that farmers are restricted to growing crops which will stand being inundated for a week or more: that excludes arable crops and modern rye-grass leys. The floods leave great tracts of land as variously rough grazing,

which during the winter time consists of sheets of very shallow water beside soggy meadows. These provide just the degree of wateriness needed by the thousands of migrant and wintering birds which use the Levels.

It is dangerous to specify that this or that moor is overwhelmingly important for birdlife: there is a tendency for conservationists to be pressured into naming the sites they *really* want to hang on to, and to accept that any other site is open to exploitation without conservation constraint. This is not a sensible corner for a conservationist to allow himself to be backed into. Nonetheless, there are places on the Somerset Levels which are vital to the bird interest, and one at least has assumed such importance that the Royal Society for the Protection of Birds has invested £650,000 in buying over 200 hectares on it. West Sedgemoor is not the only one of the very wet moors in the Levels, but it is large, self-contained, and almost certainly the highest-scoring on ecological grounds.

For years conservationists wanted West Sedgemoor to be protected and for years the relevant notifying authority, the Nature Conservancy Council, tried to get it protected on an entirely voluntary basis. There was some hope that a Somerset County Council scheme to produce an overall plan for the Levels, in consultation with all the interest groups, might provide the answer, but by 1982, the Nature Conservancy Council realized that it was too late to risk continuing with a lengthy process of consultation, and that it was time to designate West Sedgemoor a Site of Special Scientific Interest. In the meantime, since 1977, several piecemeal field-drainage schemes, grant-aided by the Ministry of Agriculture, had gone ahead. The delay did not seem to have gained much for the tax-payer or the conservationist. When finally, in 1982, the NCC declared its intention to designate more than 2,510 acres as an SSSI, there was an absurd and misguided outcry from the farming community.

In February 1983, egged on by the media, some of the farmers burnt effigies of Sir Ralph Verney, chairman of the NCC, who had in effect been sacked by the Department of the Environment for designating West Sedgemoor as an SSSI, and of a couple of the local NCC officials. They performed the stunt in the car park of a West Sedgemoor pub. Their action is doubly odd, since under the Wildlife and Countryside Act the Government is obliged to pay handsomely for the privilege of insisting that farmers continue to farm as their forefathers did on sites of acknowledged conservation value. It is rather as though the owner of a beautiful listed building had to be compensated for not knocking it down to build a factory.

West Sedgemoor is an area of around 3,000 acres, much of which is as liable to flooding as almost all the moors were until well into the nineteen-sixties. A good deal of its area properly earns the name of 'sedge' moor. In a wet winter, acre upon acre is wholly under water, while in other parts only the coarse grasses and sedges stand above the water level. Even in summer, there is occasional flooding on the moor.

West Sedgemoor is no longer the perfect Somerset Levels moorland. True, there is no such thing in what is a heavily man-influenced environment. However, some farmers have pressed ahead, with the help of tax-payers' money, in doing the sort of 'improvement' which is good for their pockets but not for conservation. Already many of the rhynes are polluted with the mustard-coloured precipitations of sulphur which occur when modern drainage ditches cut deep down into ancient peat levels. This means that West Sedgemoor, like many of the Somerset moors, looked to be going toward the soil exploitation that characterizes the fenlands of East Anglia. Previous generations of Levels farmers would never have contemplated the degree of capital investment required for this kind of work in Somerset; there was good grassland to be had here quite easily, and the climate was regarded as too wet to be really suitable for horticultural or cereal cultivation. It cannot be said too often that only Government expenditure made possible the new operations which are deeply destructive of beauty and wildlife.

Even so, West Sedgemoor is the most important site for wading birds in south west England. In winter, there are up to 10,000 lapwing on the moor, also several hundred dunlin, snipe, and golden plover, with a

A rhyne. The richness of these aquatic hedges is threatened by agricultural eutrophication.

smattering at least of widgeon and teal (sometimes several thousand when it is flooded). There are often hen harriers, peregrine, merlin, short-eared owl, buzzard, and kestrel. The Somerset Levels attract something around five per cent of the north west European Bewick's swan population, and are one of perhaps only ten sites of such importance for the species.

In spring and summer, the Levels score high as a place for breeding waders: lapwing, snipe, curlew, redshank. Whinchat and yellow

wagtail both breed here, as do mute swan (now important wherever they occur), teal and mallard. Between a quarter and a third of all redshank and snipe which breed on the generally favourable conditions of the Somerset Levels, breed on West Sedgemoor. These birds need flooded land, and some of them, like redshank, will breed in Britain in the spring in terrain which is still wet enough. The Levels were ideal for these birds in their pre-war state, but when a wetland site is drained, as most of the Somerset Levels moors have been, up to three quarters of its birdlife disappears. The moors are the most important British site for snipe, yet on one recently 'improved' wetland site here, snipe, curlew, and redshank all disappeared, while lapwing numbers were more than halved.

Lapwing. An important character for the Levels. More tolerant than some, it is nonetheless susceptible to modern agriculture.

The black-tailed godwit is perhaps one of the most sensitive indicators of the wetland health of the Somerset Levels. Less than 100 pairs of this migrant wader breed in Britain, ten per cent of them on West Sedgemoor, the only area in south west Britain where they do breed. The whimbrel is another spring passage visitor, with perhaps seventy-five per cent of the British passage population – up to 2,000 during April and May – feeding on the Levels. Redwing and fieldfare visit in large numbers, especially on Tealham and Tadham moors.

One animal provides an unfailing sign of the wildness of a riverine habitat – the otter. Its population is in great decline almost everywhere in Britain, and absent from most of southern, eastern, and midland England. The Levels were always an important place for this species. Every few years shows a further marked diminution of otter populations, even in places like the Levels which might fairly be regarded as the sort of area where man's farming activity and the otter's life-style ought to be compatible, if only human 'progressives' might be persuaded for once to stay their hand from full destructiveness. Otters need old-fashioned farmers. They need scruffy, wet ditches. The short-back-and-sides approach to waterway maintenance is hopeless for them.

West Sedgemoor, Curry Moor, and King's Sedgemoor, along with various of the Levels rivers, all provide sanctuary for this shy creature. The Somerset Trust for Nature Conservation has identified the sort of conditions essential for otters. They need high water levels in rhynes, rivers, and drains (these are canalized rivers) since this encourages fish and vegetation; they need the water to be unpolluted, for the same reason; they need rough scrub and streamside trees to provide cover and locations for holts; they need certain stretches where human disturbance is kept to an absolute minimum. They need bank maintenance and clearance to be as infrequent as possible, though frequent and sympathetic is far preferable to spasmodic and ruthless. These are all practices which once formed part and parcel with the sensible, ecologically sound and economically profitable farming which has been carried on in the Levels for the past couple of centuries. They are none of them likely to continue if misguided 'improvement' carries on.

The Somerset Levels, past and future.

Burrow Mump is a steep-sided, perfectly symmetrical little hill in the middle of a great vale south west of Glastonbury. It looks almost as though it might have been thrown up by some sculpturally-minded bulldozer-driver on his day off from remoulding the British landscape for a motorway-construction company. On its peak there is a ruined chapel, its gaping nave open to the sky, round which cattle browse. On the hillsides there are now also small saplings which may one day grow to be substantial trees.

Burrow Mump defines an important part of the Somerset Levels scenery. This is flat farmland which is almost everywhere overlooked by vantage points. Burrow Mump itself looks out in one direction over a moor which is flooded every winter and which is safer, perhaps, than many others since its owners prefer the pleasure of duck-shooting to the profit of a few extra cubic metres of the silage they could have if they drained and fertilized their grassland. There are many farmers today who believe that the Levels are best left to their own ancient devices. They believe, for instance, that the annual winter floods deposit fertilizing silts on what is otherwise very peaty soil. This may turn out to be sound in the long run. For reasons we shall come to, drainage and intensification of farming on peat soils is often very damaging in the short term and possibly ruinous in the long term.

Left. Burrow Mump. One of the lovely high places which give vantage points over the Levels' flatness.

Below. A pollarded willow in the grazing grounds of the Levels. These are amongst the few trees in the landscape, and a part of its essential character.

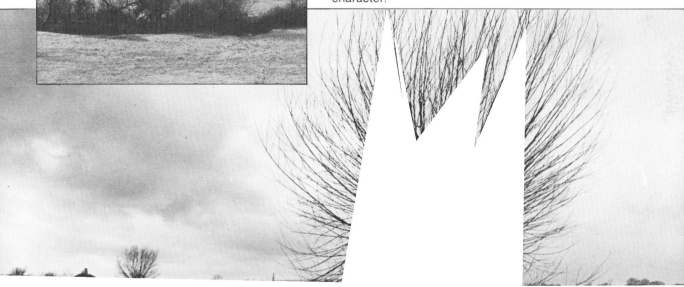

The Levels are even now a thoroughly artificial environment. There is no such thing as a 'natural' condition for these places (unless they were to return to malarial swamp), but only the degree of interference by agricultural man which seems most harmonious with the needs of wildlife. The 'desirable' picture is necessarily complicated: Bewick's swans, for instance, seem rather to enjoy grazing on improved fields which are drier than most conservationists would like to see. But what is good for the swans' vegetarian feeding habits may be much less good for the wading birds which demand a rich insect and small invertebrate life in order to thrive. We have to decide on a level of interference by man in such a place, and we have as a concomitant to decide how we should spend Government money, and apply governmental regulation to preserve aesthetic, spiritual, and prudential values in our countryside.

The Somerset Levels were formed, like so much of our land surface, by a continuous process of great and barely understood changes. Any island pokes its head more or less impertinently above the sea. Great areas of Britain are within a metre or two of the sea's surface, so any change in sea levels, due to the melting or formation of northern ice-caps, for instance, can and often has created vast flooded areas. Besides, the geology of Britain is changing even now. The east coast of England is dipping downwards into the North Sea, on an axis which runs roughly from Bristol to Humberside, and that alone threatens, ultimately, great cities like London.

Nine thousand years ago, the Levels were drier than they are now. The sea was about seventy feet lower and the area less liable to flooding than some valley areas subsequently became. Though the sea level rose dramatically, if sporadically, throughout a

West Sedgemoor from the high ground on its south west (*above*) and south east (*left*). This is the scene of the one of the bitterest conservation struggles of recent years.

3,000-year period, ending about 6,000 years ago, the river silts kept the land levels, especially the coastal edge, roughly in step with it. The belt of coastal clay (which is properly called 'levels') was being added to all the time by river silt. Thus the rivers of what was to become Somerset could not easily drain the area. The meandering rivers of the flat inland plain, pock-marked with isolated islands and fingered by the ranges of hills we call the Failands, the Mendips, and the Poldens, sauntered down to the sea through swamp-lands, then through a coastal plain much bigger than it is today with, quite likely, extensive salt marshes.

Since there are ancient raised bogs north of the Poldens, where the soil is starved of nutrients, we can guess that in places at least the water levels were constant and that there was no sudden winter flooding. But according to some authorities, even here the raised bog ceased growing properly in Romano-British times. The Levels were subject to dramatic winter flooding as the rivers, never efficient as drains, backed-up, when encroaching sea tides swept inland against the outflowing fresh water. Modern opinion has it that there were fewer massive incursions by sea-water than was once thought: they were probably mostly restricted to the seaward end of river valleys.

Sophisticated human use had already been made of the swampy areas: the oldest identified road in the world was found in the district, discovered in the peat by Ray Sweet, a peat-cutter. It runs from the raised bog of the Shapwick Heath area, towards West Heath. It is a 'corduroy' made of substantial tree branches laid down in the soggy land and since overwhelmed, and preserved, by peat. North of the Poldens, man was clearly getting some of his living from fishing and wildfowling in the swamps, and, at least in the summer, some areas might have been suitable for grazing animals. Or perhaps Neolithic man simply used the road to get hunted animals' carcasses off the swamp, or to facilitate trade in the area. Either gradually by a continuous process of accretion, or by some more dramatic event such as the incursion of the sea over the land, marine clays formed a raised basin at the Bristol Channel end of the Somerset Levels rivers, through which they had increasing difficulty in finding passage.

When the M5 was set striding across this coastal belt, the rescue archaeologists found that prehistoric man had left settlements on average every half mile or so. The Romans had built salt-workings on the coastal belt and may have contributed to agricultural

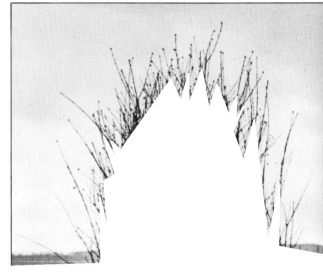

production by building sea-walls. The process of reclaiming – or drying-out – the Somerset Levels had begun in earnest. Naturally, it was slow. When King Alfred went there in the ninth century it was to seek refuge from the Danes, and Asser, Alfred's confidant and biographer and a monk from St David's, recorded that they went to Athelny, 'which is surrounded on all sides by very great swampy and impassable marshes, so that no one can approach it by any means except in punts'. Ideal for a refuge. In the thirteenth century the larger improvement schemes involved building walls round valued pasture, cutting dykes, and the beginnings of wholesale river management. There was a sharp rise in the value of land which could be kept dry enough to make wet meadows, and almost all the development which took place, certainly in the wetter parts of the river valleys – our current moors – was only to improve grazing. The great abbeys of the region, from Glastonbury and Wells to Athelny and Woodspring, were combining spiritual and agricultural endeavour on a big scale.

Even as recently as 100 years ago, and in spite of much work having been done to drain the Levels, there had been times – the winter of 1872-3 saw one – when over 100 square miles would be under water from autumn until spring. Already rivers like the Brue had clyses which allowed river water to flow out but prevented sea-water from flowing in during high tides. The work needed on the rivers was phenomenally difficult, however, and so expensive that there has never been the economic return that promoters of many drainage schemes throughout history – and especially in the last couple of centuries – have predicted.

Bundles of withy. The raw material of a basket making industry, some of it carried on by a small new wave of young craftsmen.

Nonetheless the work proceeded, especially during and after the Second World War. Britain was then in desperate need of food and many thousands of agricultural acres were pressed into service for the time being. Following the war, Britain's agricultural industry was able to argue that such a food production requirement should not have to be suddenly met ever again; it was better always to produce, however expensively, the food the nation needs. With the advent of powerful diesel pumps and also of the increasingly powerful water authorities, which always like to be seen ridding the countryside of untidy water 'problems' and love the application of large amounts of machinery and technology, the fate of wetland Somerset seemed sealed. How the argument about agriculture and investment has gone so severely adrift we will see later (Chapter 9). For the time being it is enough to say that the late-twentieth-century position is that so much improvement work has already been undertaken that there are very few hectares left which would reward improvement, and that what land there is with a conservation interest is now at a very high amenity premium.

Agricultural 'reclamation' is not the only threat Somerset's wetlands have faced, and still face. Where possible, and especially on places such as what is now Shapwick Heath where there was high-quality acid bog, there were peat-workings. Then they did not

Peat-cutting is still an important Somerset Levels industry. Households and towns could easily produce their own compost from waste. In the meantime, Somerset will be plundered.

rival the kind of work going on in the Norfolk Broads, but now peat-cutting firms, (Fisons, with sixty per cent of output, pre-eminent among them), own thousands of acres of peat which they are gradually cutting out. Little of this sedge peat is now of the right chemical mixture to make good garden humus and is therefore mixed with acid peat from Russia, Finland, and Ireland.

Unfortunately, many of the modern peat workings are on the sort of habitat which is potentially good for wildlife, and – what is worse – the peat is used to provide fibre in gardens whose owners could, with hardly any effort, compost their own vegetables. Around half a million cubic metres of Somerset's peat are extracted every year. A huge increase in recent years has led to Somerset providing around twenty per cent of the country's production. It mostly comes from an area north of the Poldens, part of which is the Shapwick Heath National Nature Reserve, where the last remains of a raised, or acid peatbog boasts bog-myrtle and ivy-leaved bell-flower, and where a summer walk will disturb the marbled white butterfly browsing the flowers.

Ashcott Heath, described as 'derelict' by one authority, though bravely lovely still, is the nearest there is now to a Somerset Levels raised bog – some twenty acres out of what once might have been 8,000 acres of this habitat. That is the order of its importance; proof that we do have a sense of our past. The past is very present in our wetlands, the places which man has not been able to tame, or on which he has not found it profitable to extinguish completely the ancient habitats he found there. They are the places where we can clearly see living memorials to the process by which the wet and the dry parts of the world came to be populated with plant-life. They have a special way of talking to us about the past. As can be seen from Chapter 7, peat is a splendid preservative, arresting the processes of decay. Trees, millennia old,

are preserved in bogs in the fenlands and on Rannoch Moor in Scotland, where a petrified forest continues to emerge slowly. Bodies, several hundreds of years old in some cases, have been hauled out of bogs in East Anglia and Scandinavia in near-perfect states of preservation. We have already seen that wooden trackways and boats can be excellently preserved in peatlands.

Peat serves a historical purpose in other ways. Just as mosses maintain their basic shape in peat, so pollen and spores which have fallen on a site over thousands of years will be preserved *in situ*. As succeeding layers of peat build up, the strata preserve a record of what pollen has drifted on to the site across the years. Radio-carbon dating is making it easier to be precise about the age of these pollen and spore remains, and thus for the past couple of decades it has been possible to analyse what has gone on in the vegetation history on and around a site. There is, of course, a degree of guessing involved in working out what plants had been on the site itself and which pollens had been brought in on the wind or by animals. Even so, a picture of the succession of plants, of the various periods of wetness or dryness in the British climate, and even of man's interference (some plants being commoner whenever man disturbs the ground), has been built up. There are also trees, like lime, whose presence is an indication of the warmth which favours them.

This is the archaeology of plants and climates. The remains of mosses found beneath glacial tills in East Yorkshire have told us that there was massive glacial activity there 18,000 years ago. It has also become clear from this sort of research that British flora, at least in the south, was never entirely extinguished by glacial activity. It has also become clear that there were variously wetter and drier periods in Britain. The successive layers of peat and clay speak of the incursions of the seas, whose levels were raised by ice-

The Somerset Levels provide wonderfully wide vistas and vanishing perspectives. This is the railway at Oath, near the pub car park where effigies of conservationists were burnt by farmers in early 1983.

melt after the last glaciation and then by a dip (though this was reversed by a later rise) in land levels. But in the peat there remains a record of succeeding generations of alder (a wet-favouring tree) or of pines and birch (generally, especially the former, dry-favouring).

Thus peatbogs have retained the record of a dry, cold, immediately post-glacial ('pre-boreal') period, which ran from around 10,000 to 9,500 years ago. This was followed by a warmer, drier ('boreal') period, until around 7,500 years ago. Then there was a warm and wet ('Atlantic') period, followed by a rather warmer and drier ('sub-boreal') period. Thus it is that, to make a large generalization, about 5,000 years ago there was the period

known as the 'climatic optimum' or 'Atlantic' period. It was in the middle of a wetland-favouring period and presaged a time when there was a setback to the bog-forming species. Just before the birth of Christ there began a ('sub-Atlantic') period in which we still live, in which our climate is fairly cold and damp, and in which it might be expected that wetland species – alder among trees, and sphagna among mosses – would do well. Actually, of course, pollution, drainage, burning, and forestry have all been far more powerful and pre-emptive influences than climate, as man makes decisions or drifts into certain habits without much regard for their ecological consequences.

However, the peat deposits of the great peatbogs do not merely tell the story of their surroundings or of general climate conditions. They also bear the imprint of the effects of climatic changes on the place itself. So Shapwick Heath, for instance, tells the story – as do the raised bogs round Morecambe Bay and elsewhere – of how reed-swamp or carr woodland was succeeded by acid-bog vegetation, especially as the drier, cooler periods gave way to the first of the climatically wetter periods around 7,500 years ago. Later still, as the temporarily warmer and drier climate set in around 5,000 years ago, there was an increase in heather cover (rather as it increased in the nineteenth century with the drying-out of our highland peatlands) and pine and birch began to colonize the edges of bogs. At the end of this period the trackways which are preserved in the peat of the Shapwick region (as well as in the Cambridge fens) were laid down, to be overwhelmed later by peat growth as the wetter climates returned in the half-millennium before Christ, and the bogs recovered their previous 'pure' status as very wet wetlands indeed.

Shapwick's end as a healthy peatbog came some time around the Romano-British period, though parts of it remained, miraculously, free from direct exploitation for peat-cutting. Ironically, Shapwick remains a last symbol of antiquity in the Somerset Levels; it is a living history book. And yet the Levels are some of the most controversial land in Britain. Almost all of them are under threat of being dried-out further, in which case they would be largely lost to conservation. Instead of a rich, complex, wildlife habitat, there would be a prairie of artificially-produced grassland with polluted rhynes.

On one argument, there is nothing wrong in farmers continuing a process which has been going on immemorially. Why should the twentieth-century farmer be stopped from improving his land as his forefathers did before him? Part of the answer comes from the requirement of the tax-payer to pay for the new 'improvement': this is a line of attack which we will pursue more thoroughly in Chapter 9. For the time being, and for those who do not want to be troubled with the complexities of the argument, it is vitally important to stress that the farmers who want to 'improve' the wetlands could not go to a bank and borrow the money to do the major work needed. The economic results of improvement simply do not financially justify the capital investment involved. Only governments could be so careless of the need to earn a profit when they invest our money.

A farm nestling beneath the hills at West Sedgemoor. The cattle are grazed on the moors, reached by 'droves'. Conservationists hope that the old pattern of farming can be maintained at least in some of the moors.

CHAPTER FOUR

The East Anglian Fenlands

The East Anglian Fens are the strangest place in Britain. The texture of a landscape so flat, so tenuously held on to by man from the sea, so hard-worked, often so mathematical, is eccentric, even if it is exciting. Its near three quarters of a million acres can sometimes be hard to bear in their flatness and uniformity, but there is lurking in these immense flat areas a wealth of fascinating detail which reveals itself to those who take the time to let the place speak and show itself to them. There is the sudden long bank of a high-walled dyke, with tall, sun-tanned grasses, and geometrical hay meadows running at forty-five degrees to the horizontal as far as the eye can see. At other places, the drainage works are flamed over with errant oil-seed-rape flowers (not to be confused with the mustard fields also found in East Anglia) running feral from nearby farms. This is of course famously rich farmland, much of it several feet below sea level. Inland, except at conservation sites, it is only the dykes and rivers, artificial or not, which provide any real wetland habitat.

The Wash provides the most dramatic meeting-place of fen and sea, and the Ouse Washes provide a marshland habitat. But the rivers and dykes are everywhere, and the overwhelmingly strong impression the place gives is that man maintains only a temporary grip over land which is always half-expecting to be returned to the sea. For people who know it only by car, it can be doubly trying. What looks on the map a short distance can turn out to be a dreary stage which goes on far longer than one thought, for this is surprisingly big country and the car's rush makes the landscape seem featureless. Contradictorily, the Fens become kinder and easier to like the longer one spends in them and the slower one goes. Foot, bike, or boat help them show themselves best. Then, the curious surroundings yield up the softness of their details. The surprises come springing out of the flat terrain.

The Fens are exciting and exhilarating, as well as weird. At Wisbech, you come across a sudden, elegant, spa-style frontage running along the canalside, like a shard of Amsterdam. From a distance, you can see a ship at berth, looking as though it had been wheeled across the landscape. At Ely, for centuries a place of refuge in a boundless, dangerous swamp, you forget altogether that this can be worrying terrain. The cathedral, looking out at grey winter days, is an extraordinary reminder of man's capacity to create; its lantern tower would lift the lowest spirits as its light colours the sun dappling down to the stone beneath. The way to arrive in this city is by boat; its watery back-door has no suburbs and no ribbon-developed roads, only reed-banked rivers running up to old wharfs and steeply-banked streets. At King's Lynn, where the churches bear marks of the great floods which perhaps explained why Noah's Flood meant so much to European minds, there are Hanseatic League houses as lasting tokens of the trading wealth of the eastern counties. A small ferry takes pedestrians and cyclists one at a time across the once-troublesome Ouse to the northern fens and the sea-wall which keeps the howling sea from its ever-threatening re-invasion.

Exactly what is included as fenland is not easy to define. The region spreads, roughly speaking, over south Lincolnshire, north Cambridgeshire, west Norfolk, and east Huntingdonshire. The Fens run north of Cambridge, east of Peterborough, and south and east of Lincoln. However, a terrain very similar to the seaward fens is also common north of the Lincolnshire Wolds towards the River Humber, and then north again on Humberside, where the sea's invasions over centuries have similarly helped build up clay deposits. These are the regions of the biggest skies in Britain. Nowhere else in the country can such an awesome, empty vista, and such devastating loneliness of scene be found so near to roads and railway lines. The land is among the hardest-worked in the country, but except perhaps for a solitary figure out

hoeing an immense field – an insect in the scale of things – there is rarely a soul to be seen.

The fenlands were formed by a complex of geological and climactic changes, but they are also uniquely the work of man, who over centuries was greedy to improve on what nature had provided. Even now, there is no certainty that man's endeavours here will prove successful. Some of the fenland is showing signs of coming to the end of its useful agricultural life. It may even be that some parts will prove too expensive to maintain, as we shall see.

The fenlands of Cambridgeshire and Lincolnshire. A staggering example of the scale of man's ambitions upon the wetlands: he has been civilising the area for agriculture for centuries. Now it seems that many of them are coming to the end of their highly profitable lives as top-rate arable reclaimed land.

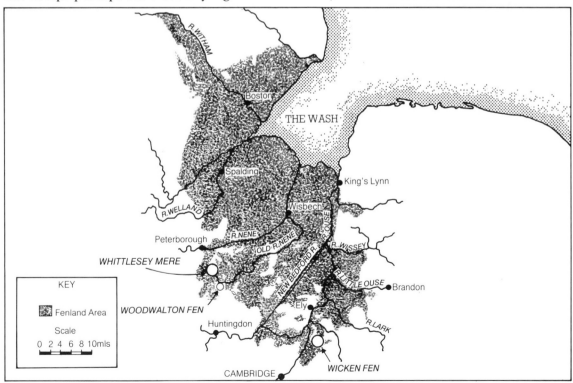

It seems likely that there was once a kind of chalk wall, roughly between Hunstanton and Skegness, which enclosed what is now the Wash. Four rivers, the Witham, Welland, Nene, and Ouse, once all tributaries of the Rhine, drained the great basin to the east of the Wash and eventually may have breached this wall on their way down to the sea which was then very much lower than now. The softer rocks inland of the breached wall were ideal candidates for erosion. The last Ice Age deposited tills and gravels over what are now the Fens. During some of this time it is possible that the Fens drained southwards through Lowestoft, since the ice in the Wash would have ponded the region and blocked the older courses of the rivers.

In the immediate post-glacial period, when the North Sea was still dry and Britain connected with the Continent, it seems that the Fens became a series of peat-forming marshes and wooded areas of pine, elm, oak, and hazel, among slow-flowing rivers. As the climate became wetter, flooding and peat-formation became more widespread, and the woodlands became fen carr or at least relinquished space to wet-loving tree species. Around 7,500 years ago the forests contained rather little elm, but much alder. There was also lime, suggesting that the climate was warm. Later, as the northern bulk of the ice-cap melted, and the sea's level rose, the rivers flooded extensively inland since their flow to the sea, always slow, was now blocked.

Away from the main Wicken Lode, there are dead straight man-made channels, providing drainage in winter, irrigation in summer, and a habitat for many aquatic plants.

Nearer the coast, the sea would have swelled into the 'valleys'. By the early Middle Ages there appears to have been a general lowering of the water levels in the North Sea relative to the British coast. This was caused by a small rise in the land levels, and merely meant that some areas – certainly those inland – became less salty, and instead of being saline and estuarine, they became fresh-water swamp.

The remains of the post-glacial trees were overwhelmed by flooding and silts or by competing peat-growth as water-tables and rain favoured different species, and then embalmed in the later silts or peat growths. They are still ploughed up today, and have been dubbed 'bog-oaks'. These hulks of great trees in a treeless landscape have sometimes been propped up, like the great bones of whales in old seafaring towns, as memorials to a very different sort of past. Within the swamp-lands there were occasional outcrops of raised ground, sometimes of older rocks, as at Ely or Boston, and sometimes where glacial deposits had given a little relief to the terrain, as at March. Stratum by stratum, a soggy-scape was being established. Clay and peat were building up in layers. In the sea-silt areas, remains of whales, seals, and walruses have been found; in the peat deposits there are the remains of wild ox, wolf, beaver, and bear.

Right. The National Trust ensure that the fields, lodes and ponds to the east of Wicken Fen are free of agricultural eutrophication. Here at least animal husbandry and wildlife are still allowed their ancient harmony.

Below. Wicken Fen lays on its brilliant dawn display.

The story of the reclamation of the Fens really begins with the Romans who, for reasons which may have been as much strategic as agricultural, set about taming this unruly swamp. They took several years to subdue the British tribes of East Anglia, and with them one of their leaders, Boudicca, Queen of the Iceni, whom they had treated very brutally several years before. They then set to work keeping the winter floods out of the swampland of the Fens. The first requirement was for scrub and wood clearance and for sea walls. 'The Britons complained that the Romans wore them out and consumed their bodies and hands in clearing the woods and embanking the Fens', wrote Tacitus. Though Roman sea-wall work remains, it seems likely that those early works of which we have evidence were at least started by the Romans. They cut a dyke which ran from Cambridge via Peterborough through to Lincoln, probably for boat-passage but also perhaps for drainage. It seems that during the Roman period the Fens were comparatively dry, though a combination of climate and disrepair brought wetter conditions during the Saxon invasions and after. It was at the eastern, seaward edges of the present fenlands that conditions were dry enough for arable farming.

In the seventh century AD, Etheldreda founded a monastic settlement at the Isle of Ely, the first of a long tradition of holy houses on the 'Island of Eels' (the usual assumption about the city's name) which, along with others in East Anglia, provided easy pickings for the Vikings in the ninth century. There is fair, if anecdotal, evidence that at least one Viking ship was dug up from the peat of Manea Fen, north east of Chatteris. John Harris, who died in 1875, told a friend that he and some relatives of his had come across a ship rising out of the peat (in fact, the drained fen was sinking) and had found her useful for kindling all that winter. "I reckon the boat belonged to some furriners wot used

to sail about these parts," he is supposed to have told his friend. The last great folklore hero of the Fens was Hereward the Wake, who used the horribly dangerous wetland surrounding Ely as a military weapon against the horsemen of William the Conqueror. In the 1920s bones of probable Saxon and Norman origin were found south of Ely.

What with silting rivers and decaying sea defences, severe flooding in the thirteenth century was hardly surprising. But there began in this period the long slow process toward the drained fenlands we now know. On the eve of the fifteenth century John Morton, Bishop of Ely, built the first of the monster dykes on which the Fens today depend for drainage. His concern was to defend the island of Ely from flooding by the Nene. However, the monastic houses, with their eye to profit and their capacity for diligence, gave way in the sixteenth century to the old chaos of individual owners protecting their land as best they could, but unable or unwilling to co-operate with one another for larger protective schemes.

The Fens were again falling prey to the emptiness and dereliction which had made them such deeply distrusted places for centuries. William Camden, writing in 1586, found there: 'a kind of people according to the nature of the place where they dwell, rude, uncivil, and envious to all others whom they call Upland Men; who stalking on high upon stilts apply their minds to grazing, fishing and fowling.' There was summer grazing, and some marsh hay to cut, and all the old pursuits of hunting and fishing.

In the seventeenth century there were many schemes, often and most famously involving the Dutch engineer Vermuyden, to drain the Fens. It was never hoped then to achieve what is common now; the greatest expectation for most of the land was that it might make rough summer grazing following a hay crop. There were many complaints against the schemes from fenmen who feared

that land that had always fed people by hunting and fishing would soon only feed animals. Oddly enough, their complaint could be echoed now by many people who dislike the modern drainage schemes, since much of what is rough-grazing land is planned to become cereal-producing acreage. The cereals are, of course, mostly for animal rather than for human consumption. The most famous protest against the drainage schemes was in rhyme:

Come brethren of the water and let us
assemble,
To treat upon this matter which makes us
quake and tremble,
For we shall rue it, if't be true, the fens be
undertaken
And where we feed in fen and reed, they'll
feed both beef and bacon.

Two hundred years later there were to be similar scenes on Otmoor, in Oxfordshire, where similar terrain was drained by farmers and landowners. Peasants saw their livings threatened, and did their Luddite best to

Wicken Fen has all kinds of glories, but amongst them there is this lode-side willow. Trees of any kind are immensely welcome in the fen-scape, and especially mature specimens.

wreck the structures – sluices and dykes – on which drainage depended. Common rights or just plain *laissez-faire* had previously favoured them with at least casual access to places rich in wildlife and sustenance for·them and their families; now they were to be ousted from earning a livelihood from newly drained land they could not afford to buy or rent. Cromwell, briefly and deceitfully, used some of the feeling against the aristocrats and the King's drainage scheme. His Ironsides, crucial to his New Model Army, drew on fenland muscle-power. Later, as Lord Protector, he himself re-employed Vermuyden to continue drainage work.

81

The two greatest diarists in the English language both made journeys to the Fens, and both hated them. For Pepys, his journey was in part over 'sad fenns, all the way observing the sad life which the people of the place – which if they be born there they do call the Breedlings of the place – do live, sometimes rowing from one spot to the next, and then wadeing'. John Evelyn complained of the stinging flies and gnats, a curse still, anywhere from the Norfolk Broads to the Highlands of Scotland, and in themselves a powerful advocate for a certain amount of cigar-smoking in the wetter parts of Britain. To complete the picture of the downside of fenland life at this time, we must note that in the early period of the drainage schemes there were plagues of rats and malaria was common. Right into the twentieth century there were fenmen who could testify that working at wetland occupations rendered a man miserably arthritic. Reed- and sedge-cutting with primitive footwear, knee-deep in cold water, must have been a desperate business. Lying on one's stomach in a wildfowling punt for hour upon hour can hardly have been better.

The end of the eighteenth century saw the development of a technique for increasing efficiency of drainage schemes in the Fens, and also a problem which meant that the whole enterprise could be lost unless the new technique worked very well. In spite of resistence from old-fashioned landowners who preferred their old wind pumps, the steam

age came to the Fens with a vengeance in the nineteenth century. The major problem was shrinkage of the fen peat. Accumulated over centuries, the sedge peat (there was very little acid-bog peat in the fens) dried out and oxidized; as it did so, much of what had been soggy and bulky became dust. The process of decay, delayed by saturation, rushed ahead. In 1851 a pillar from a building at the Great Exhibition was driven deep into Holme Fen. It eventually showed that the Fens had shrunk by thirteen feet in a little over a hundred years. The process has continued until now, when in many fenland levels there is little peat left.

In less than a hundred years it may be that these expensively engineered lands, which require constant pumping to remain free from flooding, will be poor clayland. If this is the case, they will not bear the valuable crops of carrots, celery, and the like, whose production justifies the expense of pumping. The most likely delaying tactic, an ironic one, is to keep the last two or three feet of the peat layers wet. This will require expensive new field irrigation works, and it is not certain that the system will be profitable. It may even be that the water will again win its battle against man's desire for dry farmland.

Wicken Fen

For all that most of the fenland is now intensively farmed, at Wicken Fen a patch has been preserved of the kind of worked-fen scene and system which must have been common in the eighteenth and early nineteenth centuries. It is owned by the National Trust, which makes a small charge to visitors exploring the heart of the fen, though there are free public footpaths running in some parts of it. There is a small, well-equipped visitors' centre at the Wicken Village end of the site, where displays and plant exhibits make a marvellous introduction to the history and wildlife of the place.

The wealth of Wicken's plantlife is not all extraordinary. There is plenty of the commonplace cowslip (*above*), and the lovely white and yellow lily pads (*below*), with flag iris. With the extreme shortage of natural wet habitat in the country, naturalists can no longer afford to be snobbish in awarding accolades.

Wicken Fen escaped agricultural improvement almost by chance. It was not a particularly celebrated fen at the beginning of the nineteenth century, but as the work of the fen-drainers proceeded (in the seventeenth century they had been called adventurers, because of their undertaking the financial risk, hence the name of many fens in the region) Wicken came to be a refuge for naturalists and the wildlife they studied – and plundered. The first importance of the place was that amateur insect collectors from all over the country were keen on the supplies of specimens they could find there. Early this century there was less interest in its insect life, and more in its plant-life. Now there is a tremendous vogue for bird-life, and the National Trust has responded by dredging a mere-like artificial lake which imitates the mighty meres of the Fens – some of them were several miles across – which originally helped make the region so rich in bird-life, and which were the last fen habitats to be drained. From the vantage point of a hide on the new mere, we watched, one winter's afternoon, pied wagtail, teal, lapwing, widgeon, gadwall, mallard, pochard, tufted duck, and even the atypical appearence of goosander. These are just a few of the nearly 200 bird species which breed or visit here, from the celebrated bittern to the common but magnificent heron.

A tussock of fen sedge, in glorious isolation. This is the dominant species of many fen areas, especially in slightly drier areas than would favour reed swamp. However, its appearance is often due to the management practices which made it an important part of the rural economy.

It was partly the destruction of the last natural meres in fenland which made naturalists realize that the end was coming for wild places in their wet habitat. Soham, Stretham, and Whittlesea Meres (the last big enough for storms to blow up and make sailors seasick) had all gone. The National Trust's 700 acres in and around Wicken Fen make it unique. Not only are there areas of sedge fens, great reed beds, and long, wide, mowed 'droves' (paths); there is also the richly fringed Wicken Lode· running for a mile through the site, while the farmland to the south – divided from the fen proper by the raised lode – is kept rough, wet, and free of fertilizers.

The lodes of East Anglia are arterial man-made conduits, many of them dead straight, which carry water from the uplands to the natural rivers. They pass as embanked and elevated aqueducts above the shrunken

The National Trust are keeping the fields near Wicken Fen productive but rich in wildlife. Cows leave a slightly more various sward than sheep, and they also are less likely to move *en masse* in a herd and are therefore less likely to trample birds' nests.

peat landscape to rivers like the Cam, and thence to the sea. Wicken Lode, unlike most, has a few sinuous curves in its course, suggesting that here Roman engineers were improving a natural river rather than cleaving a wholly artificial watercourse in the swamp.

Under present management, the flora of the water systems in and around Wicken Fen retains its richness. Greater spearwort, fine-leaved water dropwort, and lesser water

Wicken Fen is now a tiny island in a vast sea of intensively farmed arable land. There are now only a handful more or less untouched fen sites in the region.

plantain grow among the tall reeds and the water-lilies. The smaller dykes are rich in myriad pondweeds, water plantains, spear-worts, and bur-reed. Many of the 400 or so plant species found at Wicken are now rare in fenland. It is, of course, worth remembering that the wet habitat is now so limited in extent in Britain that we ought to be glad even of plants that have almost a weed status in soggy habitats. At Wicken there are most of the plants one hopes to find in wet places: hemp-agrimony, bedstraws, water forget-me-not, water-lilies and milfoils, yellow and purple loosestrife, fen orchid, devil's-bit scabious, marsh fern, and lady fern.

The fen proper also boasts great waving stands of sedge, reed-grass, and reeds. Each of these is achieved by cutting-regimes: they are selected by the time of the year, and regularity of harvesting. Wicken Fen is drier now than it used to be, and the National Trust is working towards making it rather wetter. Meanwhile, the tendency of most wet-land habitats to become drier simply by the accumulation of died-back vegetation and the slow accumulation of peat is accelerated by a rather lower water-table than would suit the reeds and sedges but in which the buckthorn scrub-trees do exceptionally well.

The real difference between past and present fenland is the extent to which the place is seen as a source of profit, or at least subsistence, for an agricultural workforce. A part of Wicken Fen is as it is today because the poor people never relinquished their commoners' rights. Poor's Fens were often the last to submit to improvement. It always seemed to poor people that the rich 'adventurers', with their grand improvement schemes were not giving much thought to the common people's interests.

All fen-working used a multiplicity of crops. Buckthorn, the main carr tree, might have been coppiced for twigs and small branches to use as fuel, though there would have been a much smaller area of it then than now. The fen sedge, *Cladium mariscus*, would have been used as thatching material (especially where pliability was needed) and cattle-bedding, and bundled in faggots for fuel. The reeds made, and still make, a thatching material which lasts about fifty years – twice as long as straw. The *phalaris*, or reed-grass, was grown as 'litter', or cattle bedding. The reed is cut in the winter months, especially in February and March, when, at Wicken, the great reed-stand in one field is dried out for harvesting. After the reed (whose stems are around six feet tall) has been cut, the bundles are brushed out with a rake to get rid of the shorter or more bent growths, then

Blotched monkey flower. Originally a garden escape, and now well established in northern fens and marshes.

thrown into a trestle-legged trough to be bound into conveniently-sized bundles. Each crop – reed, sedge, litter, coppice – is obtained by following a different cutting-regime in much the same sort of terrain. In nature, sedges will stand drier conditions than reeds and are therefore regarded as more likely to appear marginally inland of reeds. But for harvesting purposes, it is a question of management policy in harvesting rather than of the area in which they occur which determines which crop one will have.

Reed is cut in the winter, when its below-ground rhizomes – which work not merely as roots and the plant's means of spreading, but also as carbohydrate-storage centres – are not affected by harvesting. They continue to store the plant's life-blood for the next season. Any sedges caught up in the reed harvest will be knocked back at a time when they are least able to survive; their energy store, severely depleted at the time reeds are cut, is in their leaves. However, the sedges are harvested only every three of four years, in April or June, when any reeds in the field would be severely affected. When the sedge recovers after harvesting, it shades out the damaged reed growths. The sedge field is the richer habitat for associated plants because it is cut before other plants have come into flower: they can thus have their day after the harvest. The floristic richness of sedge fen thus depends on man's management.

After several years during which management of Wicken lapsed, with a consequent increase in scrub woodland on the fen, it is now simply a matter of ecological treatment as to whether the richness of its sedge-associated herbaceous plants – milk-parsley among them – can be developed. With a new abundance of milk-parsley, Wicken may see the return of the swallowtail butterfly, always the absent – or extremely rare – star turn of East Anglian fen habitats, which disappeared from here around 1950. Once they feel they have retrieved its habitat, Wicken's managers may re-introduce it artificially, though the 'gardening' of wildlife sites in this way is often controversial and upsets the purists.

Acre by acre the woodland is being cleared to allow, for instance, fields of reed-grass (*phalaris*) litter, with a good deal of companion purple moor grass – probably originally harvested as cattle bedding – to come through. In the autumn especially, reed-grass leaves, curling away from the stem, are the colour and shape of wood-shavings. The reed is being harvested, and the old grass trackways (droves) being mowed at just the right frequency to allow flowers to flourish. Its new managers would like the place to be a little wetter, and they would like to understand more about what harvest timings will maximize floristic variety and richness.

Many of the owners of Britain's wetland places would give their eye teeth for the wealth of Wicken Fen. It is an oddly exciting place. Perhaps it is because there are one or two mature willow trees. In East Anglia, the eye and spirit celebrate large trees as a desert traveller celebrates an oasis. Perhaps it is the softness of the place in that flat country: man-high field growths – lush grey-green in early summer, and increasingly baked and then almost desiccated later in the year – with the wind riffling through them. There is the richness of the riverine plants beside the lode, with its clear water and plump fleets of fish. It is a very favoured place.

CHAPTER FIVE

Freshwater Marshes and Wet Meadows

There are a couple of fields on the lovely small farm owned by a farmer called Tom Richards MBE, which many would regard as a bit of a liability. They lie in the bottom of a little valley near Bridgend in South Wales, and one of them in particular, is a soggy, scrubby sort of a place in which there are tussocky straggles of grass and a certain amount of soft rush. These are not signs farmers like to see in their fields. There is even a bit of a stream, which makes the ground around it virtually impassable in winter. It is an inconvenience, and of no particular value to the modern farmer who could easily pipe the stream and run a tidy water supply on his land. These fields are one small part of what we can think of as the marshland spectrum. Roughly speaking – this is a very broad definition – a marsh is usually grassy habitat which is too wet for arable farming and too dry to form fen or reed-swamp. Thus, in this country, it is almost always used for grazing and perhaps for a hay-crop. It covers a lovely and various array of soggy pasture land, rough grazing, river-fringe, and wet heath.

Not everyone loves marshland. A progressive farmer would normally take one look at Tom Richards' meadow and invite the Ministry of Agriculture advisory service to come with him for another. Between them they would probably agree that the field was ripe for drainage (though many Ministry men are becoming far more alert to conservation). Away would go the apparently unprofitable little spot, and in would come the chance to re-seed with modern, fast-growing grasses instead of the scrappy growths there at the moment. However, there would be other losses, some of which would be obvious. The unusual petty whin is found there, as are the wet-loving specialists such as bog bean, marsh marigold, lady's smock, and the heath spotted orchid. There are seventeen different species of butterfly, and myriad other insects, all of which could be expected to suffer if the field was dried out.

But there is more to the story than merely the old faked-up and over-stressed conflict between conservation and agriculture. Tom Richards is a canny farmer, glad that the Nature Conservancy Council makes him a small annual payment to keep the field as it is. However, he has noticed over the years that he has farmed there (and he has been there since 1926, man and boy) that cattle seem to like the damp little field. They do not feed there all the time, but they have free access to the field, and often wander in there *en masse* for perhaps a day out of every week or so. They do not go merely when the surrounding fields are short of grass, and it may be that they wander in out of boredom.

Some of the British freshwater marshes and wet meadows. Nobody owns the word 'marsh', so we have to some extent made our own definition. We've included some soggy bits of meadows, some river fringes – even a little pond in a housing estate.

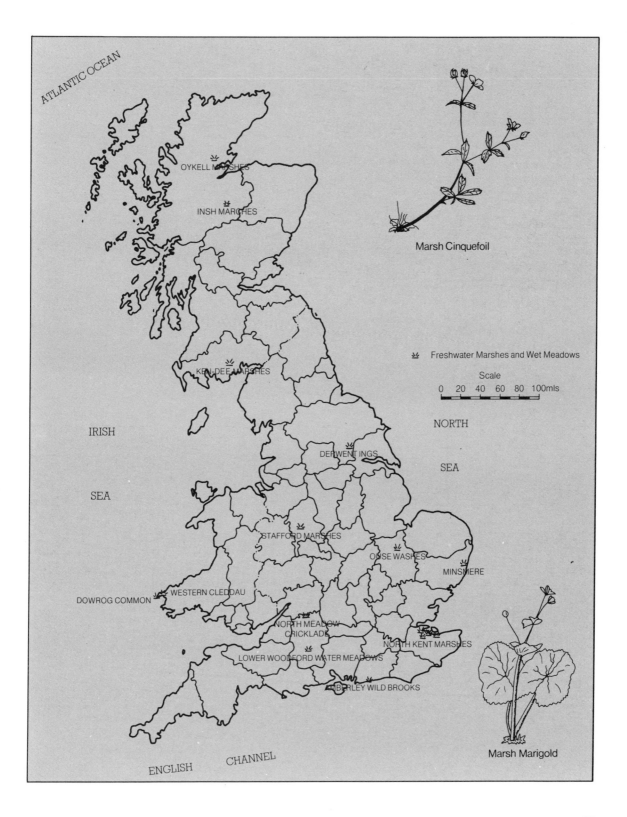

ATLANTIC OCEAN

OYKELL MARSHES

INSH MARCHES

Marsh Cinquefoil

KEN-DEE MARSHES

♨ Freshwater Marshes and Wet Meadows

Scale
0   20   40   60   80   100mls

IRISH

NORTH

SEA

DERWENT INGS

SEA

STAFFORD MARSHES

OUSE WASHES

MINSMERE

DOWROG COMMON WESTERN CLEDDAU

NORTH MEADOW
CRICKLADE

NORTH KENT MARSHES

LOWER WOODFORD WATER MEADOWS

AMBERLEY WILD BROOKS

Marsh Marigold

ENGLISH   CHANNEL

But when they get there, they graze busily. Tom Richards knows that his beasts do well at market and that they enjoy robust good health. He is inclined to think that his wet field has the sort of mineral elements in its rich variety of plants that easily make up for any deficiencies in energy-stores in its rough grass.

Tom Richards' is just one small old meadow which has never been ploughed or drained. Twenty or even ten years ago, no one would have much noticed its presence or demise. The sort of things it contains are not – or rather, were not – very rare or much prized. But in the last handful of years it has become clear that the sort of terrain represented by this meadow has been so massively diminished in area that what was once one of the commonest British landscapes has now become uncommon.

Its increasing rarity matters greatly, and in a rather special way. In recent years it has become very clear that it is not only the scientific rarity of some specimens of our plant-life which matters; it is crucial to preserve the commonality of habitats in which quite ordinary wildlife thrives and is compatible with agriculture. It is important that our conservation anxieties embrace not merely some rare orchid still growing in grand isolation for a few scientists to see and appreciate. It is also vital that there be bluebell woods, and not merely one or two but hundred upon hundred. It is a better conservation aim to insist on preserving twenty bluebell woods than to conserve three or four plant rarities.

Many of the habitats which we can loosely describe as freshwater marshes and old grassland are in this category. They include places which thirty or forty years ago were to be found everywhere, and which are now rather scarce. And they, of all habitats, should not be allowed to become scarce, for among them are the sorts of fields and commons which need to be truly common, because they are the places where one can walk freely and do no damage. They are often unfenced and there are very often paths across them. As they disappear under improvement we lose a spiritual amenity as well as a wildlife resource of scientific interest.

To return to Wales. There is a good deal of wet, 'poor' grazing pasture which is roughly synonymous with marshland. Some of it, inland of St David's Head in Dyfed, comprises a lovely little series of wet meadows at Hendre Eynon Farm whose owner, Ian Jamieson, came to realize the value of the fields he had originally bought precisely in order to drain them. They consist of around twenty acres with a little river running through them, the River Alun. The meadows at Hendre Eynon are still grazed, but they could easily be 'improved', like so much of our grassland, and grazed far more intensively, were it not for Mr Jamieson's realization that they are lovely and special. They are on a rippling terrain which produces a gently-banked warren of slopes. In May and June, the meadow is rich in yellow iris flowers, among the 160 different flowering plants which occur here. There is wavy St John's wort and royal fern by the river itself, while other, wetter parts sport four different marsh orchids (the common spotted, early marsh, southern marsh, and northern marsh) and eleven different sedges (green-ribbed, common yellow, star, glaucus, tawny, common, oval, carnation, greater tussock, flea, and greater pond). Besides the ubiquitous soft rush there are four rushes, tormentil, marsh and creeping cinquefoils, and hemp agrimony and much else.

The exquisite St David's Cathedral, Pembrokeshire, with its companions, the river and wet fields that lie with it in a small valley.

Ian Jamieson is determined that these fields should stay as they are, profitable and yet beautiful. He is proud that they are so fine, and glad to say that his cattle, turned out into the meadows for the summer, do remarkably well and seem to have fewer digestive problems than do many beasts grazed on monoculture grasslands. Certainly, they have, as well as a rich range of herbs, many more grasses than most modern animals get to bite. The Hendre Eynon range of plants runs from the abundant purple moor grass (which, occurring in association with hemp agrimony or devil's-bit scabious, characterizes many wet meadows) through to many of the grasses which appear in unfertilized and un-improved meadowland. Among them there are creeping bent, tufted hair grass, rough-stalked meadow grass, brown bent, red fescue, sheep's fescue, and mat grass, all of which, in various degrees of concentration, are impor-tant characters in the wet-meadow *dramatis personae*.

This is an area where sandy soils lead to heath-type grasses and plants. Indeed, on Dowrog and Tretio Commons which run southward of Hendre Eynon there is a won-derful and eccentric wet heathland which in some places is pure fen, in others pure heath, and in between a rich, patched mosaic of the two. This common ground was once grazed heavily, which would have encouraged the grasses at the expense of the heather and scrub; in particular, parts were grazed by horses which kept many of the pools free of the bulrushes which would now threaten to clog even the biggest of them were it not for the work of volunteers. Now much of the area still needs to be grazed, but many of the local farmers feel that there ought to be fencing, which is forbidden by the commons regulations. Such a complex place illustrates the difficulty in pinning down the 'marshland' range.

No one owns the word 'marsh' and there is no uniform usage of the word, but one can make useful stabs at eliminating what are not marshes, and move toward the kind of criteria which contribute more positively to identifying them. A fen or a peatbog, for example, is clearly not a marsh: each is clearly defined by virtue of association with large or at least crucial quantities of peat. The land which produces fen is so wet that only reed-swamp and sedge-fen manage to grow, and these go on to become permanent features, perhaps to develop carr woodland or, if they are very wet and permanently inundated by still water, sphagnum acid bog.

Marshes, however, occur on soils where there is a great deal of water, though it may be intermittent and temporary and will at least not be of constant depth. The tempo-rariness and variability of the water means that though marshland species must tolerate wetness, many are capable of surviving per-iods of dryness and perhaps of hanging on when the water flows rather fast. Thus, for instance, a riverside will often have a marsh fringe which runs up the 'shore' between the summer and winter water levels. The tempo-rariness and variability of the water on a marsh ensure that there is no chance of reed-swamp or sedge fen developing. But often, too, even when water is permanently present and free from much flow, the terrain remains free of bog development because its mineral-base is so rich that there is plenty of microbial activity and any peaty material is readily de-composed. Indeed, one way of defining

---

Dowrog Common is a wet heath area in Pembrokeshire. Very beautiful and rich in fen and marsh species, it has been getting increasing admiration from conservationists. Being common land, it was always free-range grazed by animals. Horses are particularly useful for keeping pond-choking species at bay. Now local pony trekking outfits are beginning to use it again.

marshes is by reckoning them to be areas where there is often a great deal of water, but which occur on soil or on silt. It may be the temporariness or the flow of the water which prevents the development of a peaty layer (which would take the wetland toward a fen or bog status) or it may be that the mineral-base is so rich that it rots down any dead plant material (which still produces the effect of eliminating any tendency to fen or bog).

The Nature Conservancy Council's *Nature Conservation Review* listed fourteen different major sorts of grassland communities of the kind we are mostly dealing with. Following Sir Arthur Tansley's system, they are characterized as 'neutral' to distinguish them from two other semi-natural grassland types, the grass communities found on (alkaline) chalk hills (often downland), and the acid upland grass communities of the siliceous hills (often reclaimed moorland). It is worth noting here that with the use of computers, taxonomy (the business of classifying and naming natural phenomena in grouped types) increasingly involves a kind of statistical census of plant species: for instance, the NCC's Welsh Field Unit recently undertook an attempt to classify the various major, identifiable groups of lowland wetland

The splendour of Insh Marshes. A big expanse of wet farmland that never did quite get thoroughly dried out and is now often very wet indeed.

habitats to be found in Dyfed. The process involved small sample areas which were looked at in terms of what species ocurred together and at what frequency.

The NCC's Welsh Field Survey settled on twenty-two major groups of wetland habitats, some of which broke down into up to five subdivisions of type. There is then a kind of game to do with helping to classify any vegetation mix with which one is dealing. The team devised fifty-seven divisions, or lists of plants, so arranged in two teams of either 'positive' or 'negative' indicators that one can gradually fine down to the vegetation types with which one is dealing. This amounts to a method of doing as much justice as possible to the importance of the way plants 'display' the conditions of soil and climate by their presence or absence, and predominance or otherwise.

Perhaps especially in areas of Dyfed where, for instance, Dowrog and Tretio Commons, and the straggling little rivers of the Western Cleddau system create a myriad succession of habitats, this sort of taxonomy teaches us much about the intricacies of botanical variety. It opens up the possibility of understanding a great deal about the laws which govern botanical life. It shows how this or that degree of acidity, wateriness, coldness, flow of water, or management by man plays its part in determining whether an area will boast this or that variety of fenland community, or become amenable to grazing, and with what stands of plants. This sort of knowledge matters especially with man-made or heavily man-influenced environments, as most marshes and wet meadows of course are, since they have usually been areas where grass-growths have been encouraged at the expense of reeds, sedges, or rushes. At Dowrog, for instance, it is widely believed that more grazing by animals would be useful.

So we begin to see the width of the range of wet habitats to which we can allow-

ably give the name of marsh. Some marsh will be quite dry for much of the year, while there is a good deal which is associated with larger or smaller stretches of open water beside which there will be reed-swamp or sedge-fen. Like almost all the habitats discussed in this book it is very vulnerable to human exploitation; more even than most, because the soils will usually be richer than in many wet habitats and the presence of water often much more easy to influence. The tragedy of the marsh habitats has its origin in their great potential value.

It takes as intractable a place as Insh Marshes, the immensely attractive and very wide floodplain of the River Spey in the district of Badenoch ('the drowned land') and Strathspey ('the Valley of the Spey') to survive. Agriculture has not found it profitable to reclaim them, though attempts have been made. Insh Marshes run from Loch Insh in the west toward Kingussie, five miles in length, and perhaps a mile wide. Parts of the marshes look almost African in the fading afternoon autumn light, when the huge stands of northern water sedge make a haze of beige. In drier parts of the marsh there is tufted hair grass, and in the wetter grassland areas cotton grass and purple moor grass are common. There is reed-swamp in the permanently shallow water. The water levels at Insh are anything but stable: the grassland areas have a tenuous existence as grazing fields, often being very wet even in summer. Over the years, agriculture has made determined efforts to tame Insh for its purposes, but these gradually lapsed when men appreciated the expense of trying to drain an area whose main river receives sudden, sporadic surges of water from up-country.

It has become extraordinarily rich in bird-life, and its hides, looking out over the fen and open water, are a little-known but real paradise, for the RSPB *aficionado*. There are breeding populations of widgeon, redshank, snipe, curlew, lapwing, oyster-catcher, common sandpiper, and the nearby population of the the rare wood sandpiper uses the marshes as a foodstore. Of the ducks, teal, tufted duck, and shoveler are all regular breeders here, with goosander and red-breasted merganser breeding from time to time. The spotted crake and water-rail make appearances, and though it has not been proved that the former breed, the latter sometimes do. Sedge warblers, meadow pipits, reed buntings, and a few grasshopper warblers breed on Insh.

The birds of prey are always among the most dramatic species to be seen. Insh provides prey for buzzards, hen harriers, sparrow-hawks, peregrines, kestrels, and merlins. In spring and high summer, osprey come to fish. You might hit lucky and see a golden eagle above the Monadhliath mountains, and perhaps a goshawk or a rare marsh harrier.

It takes a rare kind of luck, however, to turn up at the place late in October, as our little band did in 1982, and hang around as the afternoon mellowed and developed that special warmth which contains the threat of cold to come, and to find that the first of the whoopers have picked that very day to arrive. They came down the loch and marshes from the seaward end, a pair of them. They did not seem to have any doubt about where to come down, did not make any reconnaissance passes, but simply settled down a quarter of a mile up the marshes from the spot where we had wandered down a grassy lane to the water's edge. They did not look like creatures which had come all the way from

Iceland, these first arrivals, heralding the two hundred which would soon follow.

These are excitements of the grander sorts of marshland habitats. They are thrilling indeed, and they depend on our not always being able to tame the wetland environment. Man has always had his eye on the wet-soiled terrain of Britain. In places like Halvergate Marshes and other marshes of Norfolk and Suffolk, the need was to keep the sea out as much as possible, and then get the rain-water falling on the nearby uplands to flow through the system quickly enough to make sure flooding was at least temporary. The grassland which resulted was rich – partly because of the river silts which flooded on to them every winter – and the dykes,

Fen habitat near St David's, in Pembroke.

particularly, supported diverse aquatic wild-life.

Much of floodable Britain was in the wide, flat, valley floors of the southern rivers like the Thames, Constable's Stour in Suffolk, and especially at Sudbury, the River Arun which winds exquisitely past Arundel Castle in Sussex with the famous Amberley Wild Brooks, the navigable Nene winding among pretty towns in Northamptonshire and the Soar in the Quorn country of Leicestershire. Here the soils are the alluvial deposits of glaciers and of the rivers themselves and probably only ever needed regular grazing to create the grass meadows we see now. There was no point improving them too much; everyone accepted that the river need-

ed its overflow capacity, and that even if the water could be got off the upstream meadows more quickly, that would merely require that its stream be expensively improved down-stream.

The propensity of many rivers to flood was, however, often turned to good account by agriculturalists, at least from the seven-teenth century. Then, many flooding rivers, so far from being regarded as a threat, were encouraged to spill their banks at strategic times of year and for strategic lengths of time. The water which comes out of southern springs is often rather warmer than the ground on which they flow, especially in the early spring. There was clearly an advantage to be had from flooding meadows with a source of growth-promoting heat. A system of sluices, channels, ridges, and furrows was devised whereby water could be fed from the river, out along artificial channels, down and across the meadows, and thence into ditches which returned the water to the river. The flooding (it was often called 'drowning') was not allowed to last more than a couple of days, since many species will not flourish after drowning for longer than that. During the flooding it was essential to keep the water moving, lest rushes be encouraged.

The system seems to have worked by ensuring that the fields were free of frost, by the aeration of the soil because of the moving water, and by the influence of the fertilizing effect of the raw sewage which would have been present in rivers downstream of towns. In summer, of course, the water could be used for straightforward irrigation. In Wilt-shire, Dorset, Somerset (famously beside the

Overleaf. Where the dry farmland ends and the marsh begins. Here at the RSPB reserve, Minsmere, the waterline moves up and down several feet according to the season, creating a marginal area of typical marsh, between very wet fen and dry farmland.

North Meadow, Cricklade, Oxfordshire.
A Thameside system of ancient Lammas
meadows which are allowed to stay wet in
autumn and winter, and are used for grazing and
a hay crop. The system has never been
disturbed, and its Maytime flowering of snake's
head fritillary makes it very rare in the British
floral landscape.

Frome), and Hampshire the practice became
widespread. It was also sometimes used in
upland districts occuring as far north as Edin-
burgh. In the valley of the Piddle, in Dorset,
the practice was to flood the meadows inter-
mittently from early December until late
March. Sheep would be grazed there in the
early part of the year. After a second flooding,
a hay crop was often taken and cattle grazed
in late August and September.

The Avon, in Wiltshire and Hampshire,
is a celebrated river flowing for much of its
length on chalklands, and rich – especially
between Ringwood and Salisbury – in disused
water-meadows. At Lower Woodford, near
Salisbury, further upstream, there are old
meadows whose grass co-dominants are Yorkshire fog, perennial rye grass, and marsh
foxtail, with meadow fescue and tall fescue.
There are spring-flowering lady's smock and
the high-summer-flowering marsh ragwort,
with the water channels maintaining yellow
iris, purple loosestrife, willow-herb, water
dropwort, comfrey, and water figwort.

The Hampshire rivers and streams and
their fringes have been loved by naturalists
from Gilbert White (whose Selborne had
water-meadows) and W H Hudson through
to lesser figures such as Mrs Willingham
Rawnsley whose uncle, Hardwicke Rawnsley,

co-founded the National Trust. Her *The New Forest* (1904) is, literally, full of the joys of spring, but perhaps especially strikes a chord when she writes about the soggy fringes of the little rivers. Here she is, on a spot where horses and cattle come in the warmth of the day and congregate in the middle of a wide, open, sunny space, in a phenomenon called 'shading':

'One of the most frequented of these shading spots is called Longwater, where a stream runs between low banks, through a mile-long plain of close-cropped turf, and here you are sure to find from thirty to a hundred ponies and heifers at any time from April to October. This riband of bright green, bounded on either side by the brown and purple heather, is like a miniature Nile Valley.

'Down the centre runs the brimming stream, and as you approach it you find a fringe of golden marsh marigolds along the margin.'

This is just the sort of place where one of the eccentric valley bog systems can be found, squeezed between the fen fringe of a river and the wet grassland slopes of the valley in which it runs. This can often be proper, thorough-going sphagnum bog; the fen protects ('buffers' would be the more normal ecological term) the bog from flooding and enrichment from the river. Bog forms if the valley sides are sufficiently gently-sloped and nutrient-poor not to intrude eutrophication into the parsimonious bog-forming mosses. (This use of the term 'valley bog' is very limited. In the New Forest such spots seldom cover more than a hundred or so hectares, and they should not be confused with the huge mires and bogs which form in flood-plains associated with big rivers.) Mrs Willingham Rawnsley describes other wet places in the forest, places where the wild daffodil

crowds in damp, open grassland, and where New Forest mares, driven beyond prudence by their hunger – especially if they had foals with them – would be tempted into too-wet places in pursuit of the first new grass-shoots of spring, and would sometimes drown there. She may well have had a valley bog very near her house:

'In the beech glade, where the little clear stream deepens its channel through the soft yellow sand, and gradually undermines it from beneath the twisted roots of the beech-trees, the banks in some places are crested with thickly-growing bracken stems, in others, grown over with soft turf, and
    Here are cool mosses deep,
  And through the moss the ivies creep
while, flying backwards and forwards across the water, or poising for a moment on stems of rush or loosestrife, are the many-coloured dragonflies . . . I walked over a rough, tufted hillside, where the red heath, or bell-heather, is beginning to colour the edges of the track with its full, deep crimson, as well as the pink heath, which began to open early in June. Wafts of scent from the bog-myrtle blew across the valley of the little stream.'

Some riverside meadows, like the North Meadow at the Thames-side village of Cricklade, have been managed immemorially on a basis which allows common grazing all through the autumn and winter. After mid-February the animals are kept off the meadows at Cricklade. By May, the rare snake's head fritillary comes dashingly into its own, racing ahead in the period between grazing and the tall grass growth. Here again we are dealing with a flower whose habitat has been wrecked throughout most of its range. River improvement schemes in the past few years have meant that many once-flooding river meanders have been made permanently dry and re-seeded or ploughed-up. The fritillary is an early-flowering plant which needs to be left free to grow, ungrazed, in the early spring. The act of hay-making in July actually spreads its seed (which sets the month before). In effect, its survival depends on a hay and grazing regime such as used to apply by rivers, whose wet meadows always made the best grass. Now, cheap fertilizers have released farmers from their gratitude for the small natural advantages that wetlands used to bring; raised fertilizer prices will reverse that trend. For the time being, snake's head fritillary is now reduced to around a tenth of its presence in the 1930s, and only conservation action will preserve it.

Many grassy wetlands are still under threat of drainage, a threat which arises principally from the improvement of rivers by water authorities. In one case, however, and ironically, there was the opposite effect.

The Derwent Ings (*right*) in Yorkshire have always been a fine wetland habitat. The flooding River Derwent kept many fields very wet, and sometimes totally inundated, in the winter months. More recently, the Ings have been embroiled in controversy, though there is some hope that it has been, or may soon be, resolved.

The lovely snake's head fritillary.

The Derwent Ings ('ings' is a local word for flooding meadows) now appear to flood slightly more, and once, in 1977, they did so very damagingly, late in the spring and summer, after the water authority had built a barrage at the mouth of the River Derwent at Barmby. The greater fullness of the river is said to be impeding field drainage. The water authority had done the work because it wanted to stop salty sea water sluicing into the Derwent via the Yorkshire Ouse. Some of the local farmers, worried that their livelihood would deteriorate as a result of the surprise flooding which resulted, pressed for a pumping scheme which would save them from the difficulty in future. Conservationists were concerned that the brouhaha would provide an opportunity for land 'improvers' to dry out these ancient Yorkshire wetlands. Following negotiation, it now seems possible that after an experimental period, the position may be restored to the pre-barrage conditions.

There are now very few natural floodlands left in the country, and very few of the shallow-flooded marshes such as used once to be common on the great open flatlands of east Norfolk and on Romsey and Pevensey Marshes in south east England. They are now mostly free of floods, so that their ditches are now almost the only remnant of their wetland aspect, though they remain magnificent tracts of open countryside. The modern

A soggy Derwent Ings meadow – damp everywhere and with a flush of water in the lowest part: productive in its way, but not by modern standards.

draining of Britain has accentuated the importance of an eccentric, thirty-two kilometre stretch of 'man-made' marshland running between the parallel Vermuyden-engineered Old Bedford and New Bedford rivers. The Ouse Washes were designed to alleviate the flooding of the Ouse and surrounding fens by providing a storage reservoir. Every winter and even in some summers, these artificial rivers threaten to overflow their banks, and sluices allow water onto the land in between the Ouse Washes. There used to be several such areas in the basin of the Wash, but they have mostly now been converted to arable use. The land of the Ouse Washes lies in subtle slopes, which means that the different levels of flooding produce effects which are almost tidal, while there are various degrees of wetness, so that plants of different wet-tolerance are found in different parts of the system.

The Ouse Washlands fall into two main categories: there are areas dominated by reed

The river-and-marshland scene on the Derwent Ings was once very common indeed by British rivers. Now, support prices and agricultural 'improvement' grants have tempted farmers to change much of their land.

sweet grass and reed canary grass, where one might expect to come across marsh stitch-wort, and areas dominated by floating reed grass, marsh foxtail, and amphibious bistort. Both groups appear on alluvial soils (that is, on soils partly created by glacial, river, or sea silt), but the first is the product of greater flooding in the winter, while the second is especially open to the threat of improvement, since it rewards drainage investment by becoming ordinary, productive grassland.

The usefulness of the Ouse Washlands in the drainage scheme of things has saved them from exploitation so far, and they are now among the most highly-rated conservation sites in the country. Immense populations of birds are attracted by the 300-year-old grassland, where as many as 42,000 widgeon and 1,000 Bewick's swan have been counted. It is the nation's premier site for breeding wildfowl and waders.

To get a picture of the rarity of such grasslands, and of the conservationists' moves to do something to re-create them, we went to the Thames estuary. The North Kent Marshes were until quite recently much prone to flooding: this was classic Dickens country of swirling, dense, miserable mists. In a churchyard at St James's, Cooling, down towards the Isle of Grain on the south side of the estuary, there is a headstone with three winged cherubs' heads at the top. Beside it,

on the west, are three small 'body' stones, and to the east, ten more. The village overlooks Cooling Marshes, which are said to have inspired the opening pages of Charles Dickens' *Great Expectations*. The stones, memorials to the children of two families, mark the deaths of babies who did not survive beyond seventeen months: tragic reminders of the scourges of the nineteenth century. They have been nicknamed Pip's Graves.

Chetney Marshes, Thames estuary. Salt estuary on the left, and then the sea wall protecting the reclaimed land, with its sheep grazing.

Above. Dunlin, one of the classic wetland species. Some of those on the Thames estuary and elsewhere in winter will have spent the summer on Scotland's peatbogs.

Below. Part of Walland Marsh, in south Kent. This is now one of the few really wet places in what was once the immense south Kent and Sussex marshes. Even so, there has for many years been uncertainty about its future.

Almost any of the salt marshes in the country
could be converted at greater or lesser expense.
At Shellness (*top left*) and Chetney (*top right*)
much of the salt marshes like these have gone.
Walking at Chetney, one sees the sad hulks of
old Thames barges (*below*). Part of their trade
would have been in carting straw and hay
up-river from fresh marshes on either side
of this estuary.

In the misty marshlands of the pre-drainage flatlands, swamps were malarial places and often assumed to be choleric (which, in view of the sanitary difficulties in disposing of excreta in such places before good plumbing, was probably justified). No one much mourned their evolution to drier places, first by the enclosure of the high salt-marsh and sea-washed grasslands, and, encouraged by that first stage, the mid-nine-teenth century ditching and drainage of these man-influenced swamps. However, they now are being improved to death. Gwyn Williams, one of the conservation officers of the Royal Society for the Protection of Birds, made the area his subject for special research. Taking the North Kent Marshes as the large area of flatlands on the Swale, Medway and Thames estuaries, he has computed a loss to heavily subsidized arable farming of around a third of the area of marshes which had been there in the nineteen-thirties. Of the 14,750 hectares of grazing marsh there had been in 1935, there were only 7,675 left in 1982; some had been lost to housing and industrial development, but most had gone to arable farming, the bulk of it since 1968.

The scale of the tragedy is seen in Gwyn Williams' survey of 'dyke-nesting' birds on Chetney Marshes. On unimproved grazing-marsh dykes there were healthy populations of coot, moorhen, reed warbler, and little grebe. On a comparable arable dyke, there were only a few moorhen. Chetney Marshes are the most lovely little extent of grassland between Stangate Creek and Long Reach. There are footpaths running by the sea wall and a splendid opportunity for long blowy walks among flocks of sheep. We complete

The lovely Derwent Ings.

the final civilization of such places at great peril to ourselves, let alone the birds.

Conscious of this, the RSPB has established a reserve in the North Kent Marshes, and at Elmley Island it has leases which give it varying degrees of control over the agriculture which is carried on. To encourage bird-life, it is not only preferable to keep the grass sward 'natural' (rather than re-seed with a few species of rye grass), and important not to allow the land to be ploughed; the kind of grazing and its intensity also crucially affect the value of sites. Cattle, especially if grazed only in moderate numbers, produce a far more varied grass sward than do sheep. Grazing cattle leave tussocks of tufted hair grass and sedges which provide cover for breeding birds. Cattle are also less inclined than sheep to move *en masse*, so the risk of trampling nests is reduced.

The North Kent Marshes lost a good deal of their bird-life early in the nineteenth century, in the first great wave of land reclamation schemes, as the swamplands were turned to grazing marsh. Drainage had wiped out the avocet from here by 1850. Doing its bit to reverse the damage which has become so bad since the nineteen-sixties, the RSPB uses the dyke system to flood its own artificial mere (shallow lake), or 'scrape', on the grazing land of Spitend close to the sea wall at the Sharfleet and Wellmarsh Creeks on the north side of the Swale.

Many wintering birds need flooded or very wet land if they are to continue to visit this country, with its encouraging mild climate, from their summer countries. At Elmley, waders like to feed on the nearby mud-flats when the tide is out, and roost inland near the artificial mere betweenwhiles. Both waders (the probing birds) and wildfowl (grazing ducks, geese, and swans) need varying lengths of mostly short grasses (three geese will sometimes equal the grazing load of a sheep). Short grasses are generally more palatable and nutritious than long. The waders,

for their part, need sodden land which is rich in invertebrates and soft enough for easy probing, while many of the waterfowl in particular demand open water. The mere and the surrounding land provides that, but it also has another important element: it is kept free of intrusion by people, who are hidden from sight as they visit the mere's hide.

So it is that a February visitor to the reserve will find the hide looking out over great flocks of Brent geese, and sometimes up to 700 white-fronted geese (the second-biggest wintering flock in the country) grazing the distant marsh. The mere itself boasts big flocks of duck (perhaps up to 27,000 – so many it takes five people a day to count them) and around a thousand curlew. Dunlin and hen harriers winter here, some of them visiting from their Highland peatbog summer homes. There are sometimes up to seventy-five Bewick's swan, and up to 12,000 widgeon – a duck which often feeds in the manner of the grazing geese. Because the RSPB can keep the mere wet in spring and summer, and control the degree and type of grazing, the Elmley reserve has summer populations of breeding lapwing, redshank, and yellow wagtail, but though the snipe 'drums' here, it does not breed.

The Stafford Marshes, or Doxey Marshes, an SSSI, are a fine area of marsh and fen, with some quite deep open water where the River Sow straggles into Stafford, running close to the high-speed trains hurtling between London and Manchester, and with electricity pylons marching nearby. Partly owned by the Seven Trent Water Authority, which has been doing work on the river to improve its flow, the marshes have been embellished by little rivulets and scrapes designed by Jeremy Purseglove, who has been doing original work in making JCB hydraulic diggers friendly to the environment. Often he persuades them to make wet places yet wetter, not a common practice with river authorities. He is a landscape

Wetlands are richly eccentric. Here near St David's, grasses and sedges have grown into great hummocks.

architect with the authority, and does an important job in making sure that at least some river improvement schemes are done in such a way that wildlife flourishes after river channels are deepened. It is mostly a question of producing varied riverside environments – steep embankments in places so that kingfishers can feel safe from their rodent predators, but also very gently-sloping places which reeds and floating sweet grass will colonize, each according to its particular wetfootedness.

In the case of the Doxey Marshes this has produced wholly new dykes and a scrape

which will be uniquely useful both to birds and to the wildlife enthusiasts of Stafford. The day we were there, a pair of Bewick's swans – tantalizingly, they might have been whoopers, but we could not be sure and wished we remembered always to carry binoculars and a birdbook – were waddling uncertainly over one of the iced-over pools. The new water network at Doxey will provide the effect of tides on mud-flats by a system of sluices so that freshly exposed mud can boast the millions of worms which feed migrating birds. It will also have the gently-sloping, short-grassed surrounds to the lake which provide good grazing and a sense of safe access to the water which the wintering and breeding birds will need. While it looks a relaxed enough piece of natural occurrence, its careful design actually puts watery barriers between the walker or birdwatcher and provides some big stretches of water and

grazing where the birds will be left well alone to their own devices.

The new work will create greater expanses of many of the still-water/fen/marshland successions of habitats of plants – flowering rush and water-lily in the deeper water; reed sweet grass and bulrush on its fringes; rush-dominated grassland where the ground is very damp; and quite suddenly, as though its range had been scythed at the edges, the grazing grassland proper. By the time the work at Doxey is finished, the birds will have another major addition to the man-managed habitats which are now among the best marshland they can expect here.

We began this chapter by celebrating the commonality – as it should be – of the marshland world. It might be a good idea to end with an account of a wetland which cost £10, and which will nonetheless probably be safely preserved for ever and be visited every day by many admirers.

Jeremy Purseglove was intrigued when he came across a little patch of apparently derelict soggy ground behind the exquisite Henley-in-Arden, on the tiny River Alne. The town had always flooded from the stream which, in harmony with Jeremy's techniques of habitat-preservation (which have since been outlined in an NCC booklet – see booklist), the water authority set out to make a more reliable river. But Jeremy was fascinated by the wet waste ground which was clearly very marshy in places. It was close to what was to become a new private housing estate. He was able to persuade the planners that it would make an ideal new village pond, which was accordingly excavated by the digger-driver working on the nearby river. It was to be no ordinary village pond, but rather a shallow mere whose fringes would almost indiscernably shade into the surrounding ground.

"Within a year," says Jeremy, "at least four species of dragon-flies were here. There's the cobalt blue *enallagma*, the brown hawker, the red-coloured *sympetrum* and the pale-blue-coloured *libellula*." In late summer there will be the buttercup-yellow greater spearwort, because the digger-driver planted that and ragged robin and other plants simply by being careful to put some soil containing their seed onto the pond's fringes. They will all enliven the tufted hair grass and the reed grass.

That there should be one more pleasing English riverside pond among the houses of a midland town might at first sight seem a small enough affair. But, as researchers at Loughborough University found recently, at least half our farm ponds have disappeared this century, and in some counties over eighty per cent have gone. We can clearly celebrate any moves which bring back into our lives the tranquility and peculiar enchantment of such places.

A soggy spot in a field on Romney Marshes. This is the kind of site which often comes in for agricultural improvement, but if left as it has been for centuries, can be quite profitable, whilst providing a habitat for birds, and – with the right regime of harvesting – lovely flowers too.

CHAPTER SIX

The Urban Wetlands

About four miles from the National Westminster Bank's soaring tower in the City of London there is wetland of curious, almost furtive, charm. It is called Walthamstow Marshes, and its career as a place of first agricultural and now natural value has been chequered. It remains, however, a great boon for this part of the world. One February morning, my elder daughter and I sat on an icy hummock waiting for Glyn Satterley to set up some photographs in this canalside miniature wilderness in north east London. We watched the clattering trains queue up for entry to Liverpool Street station, and waved at the passengers and got tooted at by the drivers. This is no isolated, privileged spot. One of its greatest admirers, the self-taught botanist Brian Wurzell, once described the burnt-out car hulks which litter it as 'Walthamstow's own special fauna'.

Right. Some Urban Wetlands. In urban sites up and down the country there are wetland habitats that have survived to delight children and walkers. Now they are being fiercely defended.

Walthamstow Marshes (*below*). Railway lines may appear to be odd companions for wildlife. But if the wildlife isn't too fussy, why should we be? At Walthamstow Marshes stands of reeds and sedge beds are a delight.

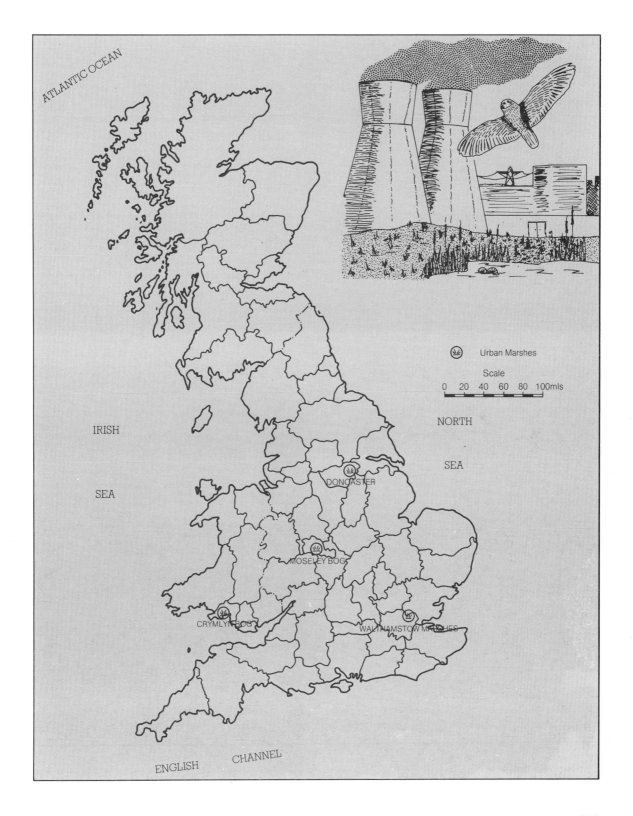

ATLANTIC OCEAN

IRISH

SEA

SEA

NORTH

SEA

Urban Marshes

Scale

0   20   40   60   80   100mls

DONCASTER

MOSELEY BOG

CRYMLYN BOG

WALTHAMSTOW MARSHES

ENGLISH   CHANNEL

Above. A short-eared owl. They can be seen hunting over Walthamstow Marshes (*right*), even so close to blocks of flats nearby, and in spite of constant 'intrusion' by walkers. Walthamstow boasts good stands of reed mace, amongst much else.

As a boisterous three-year-old chattered and called out across the marsh's frostiness that morning, her noisiness did not deter a short-eared owl from finding its lunch. The bird cruised in the laziest, most sparing, yet diligent, manner across its terrain: back over the embankment, out of sight, into the marsh again, a dalliance over the sedge beds. Periodically the bird – large in its leisurely, understated way – would drop down and be lost for a moment in the tall grass, but either we saw too little or it was finding the prey thin on the hoary ground, for it would soon be airborne again and on the hunt. It ignored us altogether and came quite close very often, as startling in its proximity as an airliner at roof-top height.

That this bird should come so close suggests all sorts of things about Walthamstow Marshes. One, of course, is that even these eighty-eight acres so near the heart of London can provide a food store for a big bird of prey. There is also the thought that in spite of getting a living on a haunt for kids not usually known for their benign instincts where wildlife is concerned, this short-eared owl had not been harrassed into anything like a mortal dread of all humans. Perhaps it knows how to tell the look of an admirer from that of an aggressor. Whichever the case, we felt mightily privileged, as indeed anyone might who came to this marsh.

However, the charms of the Walthamstow Marshes were not always much advertised or understood. It took the determined efforts of several local people, led by John and Jane Nash and Brian Wurzell of the Save the Marshes Campaign, first to discover, then chart, and fight for the survival of the site. It is fair to say that many urban wildlife sites have not been much recognized by the orthodox authorities or their experts until very recently. The newer awareness that wildlife sites carry a premium if they are near dense populations has mostly come about because of the enthusiasms of local campaigners.

Walthamstow Marshes are not spectacular enough, nor positioned in so grandiose a landscape, to attract the attention of the older school of wildlife protectors. But for all that they have been under-appreciated; the Marshes are handsome. The reed-beds are extensive and accessible and as dramatic in their way as those at Wicken Fen (perhaps more so for being unexpected), and the large central areas of the site, especially when every wellington-boot-step is a risky one, are rich in sedges. In summer, the meadow-sweet, great willow-herb, yellow flag, and golden rod, are in flower.

Part of the good news is that the Michaelmas daisy is also in flower from August to September. It is also part of the bad news for Walthamstow Marshes that it should boast Michaelmas daisy. For purists, one of the criteria by which a site can be declared really valuable is that it should be free of signs of disturbance by man, and Michaelmas daisy is a clear sign that Walthamstow Marshes have been disturbed. Indeed, the area was partly dug over with defensive trenches in the Second World War, for instance. Part of it was the scene of the first all-British powered flight by A V Roe in 1909. It has been used as a fairground in its time. Nonetheless, this and one other site in the Lea Valley (Rye Meads) are all that remain of the very soggy meadows which once wallowed in the Lea's flood plain.

It is likely that the Lea was one of the first rivers in the country to be canalized: it was too useful for navigation, its fringes too useful for agriculture, to be left in its primitive, floodable, and dangerous state. But it was never tamed, and its fringes were always sodden.

The marshes at Walthamstow became Lammas-land, with parishoners of the then embryonic Walthamstow having grazing rights on it. The original Walthamstow Marshes comprised about 140 acres and were divided into 130 strips of variable size. Horses and cows belonging to the villagers were allowed to graze from old Lammas Day (13 August) to old Lady Day (6 April): during the hay- or corn-growing season (known as the defence period) the strips reverted to the exclusive use of those who owned them. The system was certainly operational by the fifteenth century, and only ceased in the twentieth. There was a prolonged struggle over the passing of the old customs and ownership patterns on many Lammas-lands in the country, but not at Walthamstow, where the coming of the railway and the gradual erosion of the farming world in east London put paid to the old ways. By the 'thirties these ancient meadows were used only for recreation, the fair, a little boating business, and perhaps some hare-coursing.

In the 'thirties the local borough took the Marshes into public ownership. Eventually, the Lea Valley Regional Park Authority took part of the ancient place under its ownership, paying a premium for the gravel left behind by the glacier which had scraped the place flat in the first place. Then began the great battle for the site. The Authority felt it was obliged to reclaim the gravel and then turn the site over to traditional recreation space, with boating lakes and football pitches. Enter the Save The Marshes Campaign, which successfully fought for the wildlife interest in the site, and finally persuaded the Greater London Council to refuse planning permission. Since then there has been a quiet further campaign by the local fans of the Marshes to have them officially rated and defended for their wildlife value.

At the time of writing (March 1983) it is uncertain whether Walthamstow Marshes will be accorded the highest sort of protection allowed to a site in Britain, and be designated a Site of Special Scientific Interest by the Nature Conservancy Council. One school of thought is inclined to say that though it is interesting that such a place should be found so near the centre of London, it betrays too

much of its muddled past to rate the highest scientific interest. Purists like a site to be free from invasion by plants 'alien' to the kind of habitat it typifies, so as there are escapees from London gardens at Walthamstow, the place goes down a notch or two in many people's minds.

There are some absentees which it would be reassuring to see in such a wetland site, and there are other plants whose occurrence rather militates against its being pure. Yet it is precisely the presence of Michaelmas daisy and comfrey – both of them to some extent signs of disturbance – that encourages some of the many insect species to thrive on the site. These two plant species complement one another by making useful nectar at just the right times of year. Nature is not as fussy as scientists sometimes are in assessing the value of a piece of habitat.

Ultimately, the site's very diversity will probably argue for a higher rather than a lower estimation of its worth: to have 350 different species in such a compact little mosaic so near the heart of London would be worth preserving simply as an example of creation's capacity for serendipity. The presence of sneezewort and adder's tongue fern indicates that the place has escaped agricultural 'improvement', since these plants are very susceptible to damage. Burnet saxifrage and crested dog's tail, both present at Walthamstow, indicate old grassland, while brown sedge, lesser reed mace, water dropwort, and meadow-sweet are valuable so near London. There is also an astonishing range of hybrids, including seven hybrid docks.

What is clearly required is that the place must be accorded some sort of nature reserve status. Some proper balance must be struck between the site's need for some conservation management (it badly needs grazing in places) and the strongly felt need of many local people that it be accessible to them for walking in the nearest thing to country the area can boast. More than that: messed about though it has been, it is actually far more natural than most farmland in the nation's countryside. It just happens to be set in a landscape of embankments, reservoirs, and housing estates, rather than of rolling downland.

Yet more complicated is a wetland site in the middle of Birmingham. Moseley Bog is the most dense and compact little wetland imaginable; within a few acres are a streamside, a wet woodland, and a patch of marshland. It had come to be regarded as wasteland and part of it was about to be built on by housing developers when some local campaigners took its part and called a halt, insisting that their childhood haunt deserved better. It appears to have been a hang-out of the young J R R Tolkien, who is said to have used it as a model for some of the scenes in *The Lord of the Rings*.

Part of the site is made up of Victorian gardens which have run wild, part by erstwhile mill ponds which became redundant and whose dams were broken down. The marshes are an amalgam of ancient and modern, with prehistoric camp-sites occurring beside the Cold Bath Brook (pressed into service as a mill stream by the eighteenth century) and derelict Victorian gardens. Yet it is this site which boasts wood horsetail and the uncommon royal fern. Woodpeckers clattered in the quiet of the morning when we visited. In spring, the woody ground is a misty blue scatter of bluebells.

The Soar River at Ratcliffe-on-Soar, south of Nottingham, whose banks are much used by fishermen and walkers from the nearby cities and towns. This is a controversial river because it has for years been the subject of debate as to whether its channel should be deepened in places to make it flow somewhat faster. Drainage in nearby fields could be improved following such a scheme and the grazing meadows nearby 'improved' for agriculture.

The modern story of Moseley Bog is a further instance of the need for local enthusiasts. It was Joy Fifer, who lives in the area, who first kicked up a fuss when she heard that Moseley Bog was to be developed for housing. It was her initiative, and that of other local residents, which persuaded local councillors to plod through the mire and sense the place for themselves, and which got the site the kind of reputation which made it hard for the developers to continue with their original scheme. Now it looks as though the site will be preserved; some developments are likely, but not the sort which will spoil the place's essential value.

Out of the campaign for Moseley Bog has grown the Urban Wildlife Group which, though Birmingham-based, became the seed for other similar groups and which is still vigorously fighting for wildlife sites in cities. A campaign of particular importance in the Birmingham and wetland contexts is the one for the River Rea, nicknamed 'the Mother of Birmingham', a potentially lovely stream which has been devastated, quite unnecessarily, in the name of better drainage.

Another classic case of urban wetland preservation which will certainly go down in conservation history is that of Crymlyn Bog at Swansea, South Wales. This fen and bog site was mostly owned by the Central Electricity Generating Board and was in danger of becoming a rubbish-dump. It took a campaign by the local Friends of the Earth to stiffen the resolve of the conservation authorities. Only when they issued a writ against the Nature Conservancy Council for not working to protect the site more vigorously did the authorities seriously get down to

business on its behalf. Until the pressure groups changed the position, the orthodox authorities seemed, as is all too often the case, simply to protect the most obviously valuable parts of wildlife sites. It looked as though they were prepared to see some of the site go under in a compromise which would, for example, have lost up to fifty acres (of the 500-acre total) in which there occurred the uncommon marsh helleborine and marsh arrow grass.

Near right. Moseley Bog, and a wood horsetail, an uncommon plant in Britain, and one with a very ancient provenance. *Far right.* Wet-footed glade at Moseley, a spot which demonstrates fine wetland habitat surviving even in the middle of cities.

Where there is hardly any already-existing wetland beauty to protect, it can sometimes be encouraged almost from scratch. Within sight of London's Tower Bridge, for instance, there is the William Curtis Ecological Park. Some 350 lorry loads of topsoil and the efforts of dozens of volunteers over five weekends (and more subsequently, of course) have transformed a Thames-side wharf site into a place where a little lake is reed-fringed, and where bog bean, marsh marigold, and yellow flag all flourish. We have so successfully banished marshland from our urban lives that it now appeals to us to recreate it on a Thames-side spot. It behoves us to remember that many of our big cities were built beside marshes – as could be expected, since big settlements usually accrete round some estuarine access point to the sea and the world beyond.

It is a pleasing thought for those who like to dwell on the vagaries of life that when a rich family called Grosvenor had the good sense to commission Thomas Cubitt to develop a soggy marshland site they owned – it would be called Belgravia – as a housing estate in the first part of the nineteenth century, their family fortunes rose to such heights that they were rewarded with a dukedom, and became the Westminsters. Thomas Cubitt promptly hired 1,000 men and used the soil he was excavating from St Katharine's Dock (much of dockland, incidentally, had been marshland) to create something dry and profitable where there had only been swampy meadow before.

People have been making money from drying-out wetlands for a very long time. That is why what we have left needs to be protected very carefully. We may not now appreciate the full worth of these places, but that is because it has not yet impinged on the public consciousness just how few acres of them are left, and just how lovely they can be.

Luckily, they have their friends, who must defend whatever sites they find, and wherever they occur. Oddly enough, the nineteenth-century railway engineers often unconsciously worked as excellent conservators of habitat: their curves and triangles and embankments often (as at Walthamstow) effectively locked valuable sites up and kept them safe from building work. South of Doncaster, the 250 acres of Potteric Carr, hemmed in by railway lines, have been saved and improved by conservationists working in close harmony with British Rail, whose landholdings probably include more delicious little wetlands than anyone else's.

The rushing scene outside a car or train window is as likely to hold wetland value as is the country ramble. That being so, it is a sound idea for us to embrace the notion that seeking natural loveliness does not depend on doing so in rural solitude. After all, if that East End short-eared owl is content to hunt its prey surrounded by trains and people, then we should be prepared to see beauty wherever we may, even right under our noses.

Crymlyn Bog, Swansea. This is now one of the most famous urban wetland sites, and one which demonstrates how wildness near cities is threatened by industrial development and use as rubbish tips. It took the concerted efforts of local enthusiasts to put the place firmly onto the agenda of conservation concern with the orthodox authorities.

CHAPTER SEVEN

The Acid Bogs

What the great ecologist Sir Arthur (AG) Tansley called 'Britain's green mantle' is now overwhelmingly a patchwork of fields and forest with a scattering of heath and moorlands, and occasional variations, usually at the fringe of land and water, in estuaries, cliffs, and beaches, or beside watercourses. Yet this view of the country belies the million and a half hectares (around a twentieth of the total land surface of the United Kingdom) which are covered by peatland, even though much of it is now intensively farmed, under forestry or otherwise severely damaged. It is a modern picture, and not complete even today. Wetlands have survived in surprising places.

We have already looked at some fens and marshes. In the wet west and north, there is another sort of wetland scenery altogether: it is wilder, and man's hand seems less obvious. Though much of it looks like moorland, and human activity helped it become so, there are still pockets, sometimes awesomely large, where its ancient character remains relatively untouched. There are great empty stretches in the Pennines, throughout Scotland, and in Wales and the Lake District, where sheep, shepherds, and rucksacked walkers are the only obvious sign of human activity. These areas are often dominated by moor species, with heather, deer grass, and purple moor grass as the main vegetation for mile upon windswept mile. In these regions, where you can now walk dry-footed most of the year, there was once an immense amount of living, soggy, peat-forming bog. There was peatbog over much of wet lowland Britain, and there

was thin but vigorous bog growth on many of our wet uplands. And for those prepared to leave the beaten track, Britain still maintains a richly complex, exciting wetland habitat which is immemorially ancient and primitive.

It is appropriate that the most ancient and primitive of our plants – and among wetland species there are many of both – should be found in these eerie and awesome surroundings. Thousands of acres of the oddest landscape in Britain are composed of peatbogs. There is little immediate or obvious charm to many of them; they do not invite the eye of the casual observer, but they are hauntingly splendid and reward attention. To love them is to be a part of a small but growing fanclub. To the extent that they remain lonely and largely misunderstood, even feared, they represent a kind of secret world, though it is one to which anyone can

---

Some of the Acid Bogs of Britain. We travelled mostly amongst the peatbogs of Scotland, which are the largest of those remaining. All the growing peatbogs shown here are much smaller in extent than they would once have been, some by a very great deal. This is amongst the most diminished habitats in our islands.

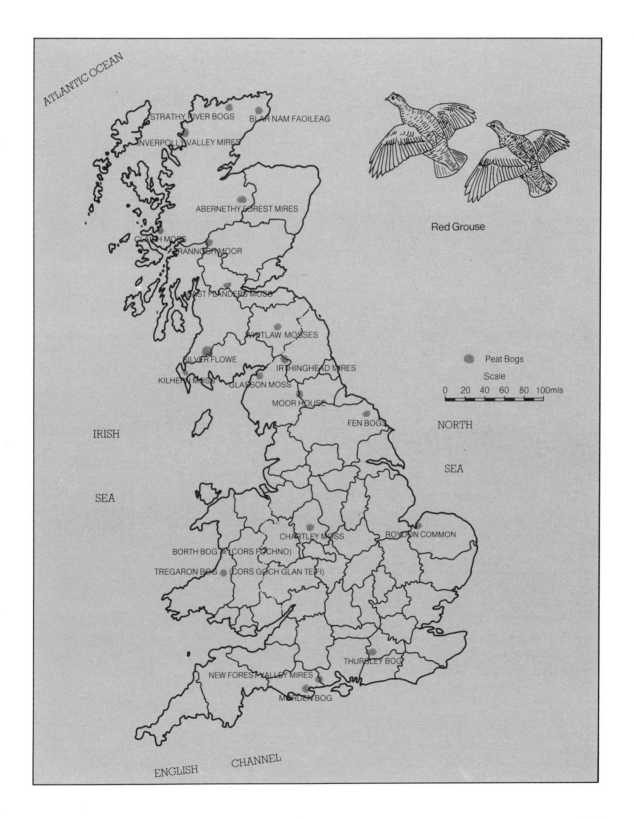

ATLANTIC OCEAN

STRATHY RIVER BOGS
BLAR NAM FAOILEAG
INVERPOLLY VALLEY MIRES

ABERNETHY FOREST MIRES

Red Grouse

CLAISH MOSS
RANNOCH MOOR

EAST FLANDERS MOSS

WHITLAW MOSSES

SILVER FLOWE
IRTHINGHEAD MIRES
KILHERN MOSS
GLASSON MOSS
MOOR HOUSE

FEN BOGS

Peat Bogs
Scale
0 20 40 60 80 100mls

IRISH

NORTH

SEA

SEA

CHARTLEY MOSS
ROYDON COMMON

BORTH BOG (Y CORS FOCHNO)
TREGARON BOG (CORS GOCH GLAN TEIFI)

THURSLEY BOG

NEW FOREST VALLEY MIRES
MORDEN BOG

ENGLISH CHANNEL

gain physical access. Once you are on bog, the keys to enjoying it are an informed guide and an eye and mind educated to the virtues of the vast flatness or gentle slope which is not merely a characteristic of the way bogs look, but a crucial defining characteristic of the kind of place where they can be found. In my case, I was lucky to have the fun of travelling through much of Scotland's peatlands with Richard Lindsay, the Nature Conservancy Council's peatbog specialist, and a man whose enthusiasm was infectious. He had the ability to see the wood for the trees – or in this case, the landscape for the moss.

Bog can be formed anywhere in the whole range of the British climate: but bog crucially does not mind coldness, and it needs to be fed by water which is neither enriched nor polluted: only rain will do. Unlike other wetlands, bogs are, ironically, not fed by ground water at all. They can grow on slopes as steep as fifteen degrees and, if the climate is very wet, even steeper. A peculiarity of the mosses which form bogs is that, though they like to be thoroughly saturated, they will tolerate no movement in their watery habitat. The wind-ripples on a large open pool are enough to halt their growth.

Below. A grouse moor in Sutherland. It looks wet. It *is* wet. But it is not growing sphagnum mosses any more. Lovely, nonetheless, with its own special wilderness charm and many of the wetland species.

Above. An Argocat in action. This is one of the few vehicles that can make any progress in wetland sites.

The sphagnum bog-mosses, whose presence alone defines bog from marsh, wet moorland, wet heath, fen, or any other wetland, are a delicious mixture of the extremely fussy and extremely hardy. They are also very beautiful. Like much of nature, they will stand only a little exploitation by man. However, there is fierce debate about what should be the future of vast tracts of their habitat. Part of the debate's greatest force stems from the fact that man has already, to a greater or lesser extent, damaged almost all of the bog in Britain. It now only remains to decide whether we should be trying to reconstitute and repair much of the damage, or continuing along the path of exploitation, and 'improving' thousands of hectares in what will ultimately prove to be the destruction of one of our last wilderness tracts.

Peatbog is the natural habitat on almost all flat or gently sloping land of poor drainage in the British Isles, at least during those thousand-year spells when the climate is relatively wet, as it has been for the past 7,000 years. It would once have been found in vast quantities, and those places where there was – 7,000 years ago – no bog, would have been poor mountain scrub or more or less forested. This picture of huge tracts of soggy peatbog in areas scraped flat by the work of glaciers until the end of the last Ice Age about 10,000 years ago would have occurred in the wet valley floors of the Severn and what are now the Somerset Levels; it would have been a main constituent of huge areas of East Anglia, and would have included what is now the rich farmland of Lancashire, Lincolnshire, and Cambridgeshire. Now we have largely banished this peatbog to a few last outposts where it not merely *can* grow, but is virtually the only thing that *will* grow. These are the classic 'raised' and 'blanket' bogs. There are isolated cases of 'valley' and 'basin' bog.

Blanket bog is the poorer, thinner cousin of raised bog. The full luxuriance of raised bog in effect spawns its slimmer relative which is mostly found on Britain's uplands, from the mountains of Wales to the Pennine Chain and the great Scottish hills. What we have come to think of as moorlands, where heather predominates, once mostly supported peat-forming mosses which spread across the high mountain plateaux as blanket bog. It is uncertain whether the upland peat-producing bog followed what had been more or less exclusively scrub and tree cover, or if there was a complex mosaic of the two. Certainly, it happened that man and the weather simultaneously gave the peat-forming mosses their chance. Between 5,000 and 7,000 years ago, man cleared much forest cover for food, fuel, and shelter at about the same time that a wetter climate set in. Mosses will stand, indeed demand, greater wet than forest, and are capable of overwhelming forests; but they might, historically, have done all the better in places where the trees had been plundered already and there was diminished competition.

Whatever preceded the blanket bog on our uplands, it began forming, and thus had its genetic origins, in the raised bog which formed in an extraordinary succession on many wet, flat places. Raised bog can be thirty, forty and fifty feet deep. One can walk on it for hours, and all the time be house-high above the ground. Blanket bog is a kind of huge carpet of peat mosses colonizing from raised bog. It is a great continuous outpost, usually on more or less unpromising, quicker-draining ground where a bog could not have formed by itself, spreading out from what one could think of as parent bog. Blanket bog amounts to a case of the mosses themselves carrying wetness out over dry habitat from places which were geologically wet originally.

The landscape of Britain gives a solid account of several periods of glaciation. The mountains and valleys are the story of ice imprinting itself, mostly through the scouring effect of the rocks it carried along, on the terrain. Thus even the most solid features of our geology are the result of weather. The glaciers made the valley floors and the great plains, and coated them with impervious clays. This geological phenomenon created the conditions for colonization by damp-loving plants and eventually by the peat-forming mosses. Succeeding warming and cooling of the northern hemisphere effects the story of the wetlands because the alternating periods bring with them either greater or lesser wetness. In the broadest possible terms, wet periods favour mosses and dry periods favour trees. Across wet millennia, the British landscape will support vigorous moss-growth. When the climate is drier, scrub and trees will do better.

It is rare that one sees *racomitrium* (like many mosses, it has no English, or common, name) growing on anything but a damaged bog. But Claish Moss, Argyll, is a healthy bog and has always grown more than its fair share of this beautiful, bright moss in its distinctive hummocks.

Kilhern, Dumfriesshire

A search for the great, primitive unspoiled British peatbog could have no better beginning than Kilhern Moss at Glenluce, in Dumfries and Galloway. The entire area is called the Moors, and through it runs the valley of the Water of Luce. A small road takes one past the lovely ruined Glenluce Abbey. A stony track bounces up the valley side beside a busy subsidiary stream whose fringes support bog pondweed, marsh arrow grass, and lady's smock. March up it, through terrain which has scattered outcrops of rock and heathland (it is too steep for proper moss growth), among tufts of deer grass, and suddenly the footing changes dramatically.

Kilhern Moss is on top of a hill, but it is a hill which happens to form a broad, shallow crater-shape: it is a kind of dimple perched on high. 10,000 years ago, the speculation goes, there was a lake in this dimple. But now, instead of the lake in the bottom of the hollow, there is a peatbog which has filled it right up. It may be forty feet at its deepest. It seems almost to threaten to overflow the geological features which have so far contained it. It is growing ever upward, making a gently sloping dome which at its centre is much higher than the walls which are holding it in. It is only the constant compression of the dead mosses below the living surface layer that allows the bog to grow without spilling over; generation upon generation of moss presses down on its ancestors.

We can suppose that the geological dimple was formed by glacier scouring, but the genesis of the bog can only be guessed at. If there was simply barren rock which was wet but not flooded, lichens would have colonized it. In a slow and gradual process, they would have become minuscule traps of plant litter, which itself would have provided the opportunities for non-sphagnum mosses. Once they gained a foothold, they would provide the stable, very wet conditions required by sphagnum, bog-forming mosses. If, however, the glacier's action formed a static catchment area for water in a rocky-bottomed lake, then, providing the water was acid, sphagnum mosses might have formed a kind of raft, gradually growing out from the

---

Claish Moss is rich in *racomitrium* and in the classic pools which distinguish bogs in their high state of development.

edges of the lake until it formed an expanding dome over the water, a 'basin mire', and there was the ideal soggy, acid environment for the growth of a mature bog.

The majority of bogs form in the classic reed-swamp/fen/bog succession. Had the mineral base of the area been rich itself, that would have made the water running through or trapped within it slightly rich. It would thus have been ideal for reed-swamp formation. There may have been many different sorts of plant in this wet lake-fringe. Certainly, over millennia, the constant accumulation of decayed reed-swamp communities would have created peat which would have gradually encroached on the lake, making it smaller and shallower. Ultimately, there

would have been no lake at all, but a soggy, peaty morass. Now, the top layer of plant-life growing on the accumulated remains of this gigantic compost heap finds itself perched slightly above the water-level of the surrounding land. There was probably no real quantity of open water, and no contact with the ground's alkaline minerals.

A peculiar position was reached perhaps 5,000 years ago. Ironically, this slow, steady development ultimately meant that conditions had been created which were ideal for one of the world's more primitive plant-forms to thrive. Sphagnum moss, of which there are many distinct species, is the great 'acid' peatbog former. It is primitive in the important sense that, as a bryophyte, it has no roots and no system for pumping water and nutrients through its structure: it is not a 'vascular' plant. The bog-forming sphagnum mosses cannot receive nutrient from the ground below. They cannot survive movement in the water they love. They thrive only in environments which are very lean in those things which promote the growth of the higher plants. They like water, plain and simple, un-enriched and static. Moving water passes over different rocks and minerals; it brushes against various vegetation forms; as it does so, it picks up nutrients which kill bog-forming sphagna.

If nature could appreciate stories, it would be very keen on the tale of Jack Sprat and his wife. There are plants which will not live high on the hog; they will not tolerate the fat of the land. One such is sphagnum. The kind of environment which would kill most plants by being too poor in nutrient might kill sphagnum by being too rich. Sphagnum can get its living from rain water which, as if by a miracle, contains all the food it needs. It is part of sphagnum's modern tragedy that there is too little pure rain water around these days for it to do as well as it might. At least in much of Scotland there is plenty of rain: the western regions can get more than ninety inches a year, and between forty and sixty inches, rather rare outside the far west in England, is normal in much of Scotland and Wales.

This primitive plant is a born aquatic: it builds itself not on soil, but on water and on other members of its own tribe. The processes – oxygenation through water movement, or liming – by which nutrient is released from dead vegetable material was partially arrested on Kilhern, as it is in any bog. As the vegetation partially decayed, it released humic acids which increased the acidity of the vegetable mass. This helped to slow further decay and provided further advantages to bog-forming sphagna. In the acid, soggy environment there was neither oxygen nor nutrient to allow the decomposition of old generations of peat by microbic life-forms. As sphagnum mosses colonized the swampy, stable peat-and-water world of the lake in the dimple on the hill, they grew and spread and died, but they did not decompose. The dead sphagnum was pressed down into the watery world below the living layer, and could not decompose. It is running water, air, or animal life which helps ordinary vegetable matter decay in the soil. In an acid peat there are none of these, so there is very little decomposition.

Thus, up on Kilhern Moss, there is almost a square mile of peatbog, up to twenty or thirty feet deep, which quietly grew at the rate of about two millimetres a year. There is on Kilhern a living layer of sphagnum perhaps finger deep. Below that there is more or less soggy peat, very wet at the top and saturated at the bottom. The sheer weight of 5,000 or 6,000 years' worth of peat makes sure that the lower layers are impermeable. The rain that falls on the peatbog takes an age to drain through it, and indeed more is lost through evaporation than through drainage, in a classic, undisturbed bog.

In many ways, Kilhern is a classic bog. Most obviously, perhaps, it has the required

wetness. Every footfall is a squelch in such a place, where ordinary wellington boots are the least of what is needed. If you stand still, you find a pool of water has spread at your feet. You cannot sit down, because you will create a puddle for yourself. The plant's water content is immense. You can bend down and pick up a handful of the growth at your feet and wring it through. It is like wringing out a bathroom sponge, except that the plant will spring beautifully back to shape afterwards, and its shape is richly branched, and composed of individual plants like plump pipe-cleaners. Wetness is not an infallible sign of health in a bog. but it is a good beginning. Kilhern is very much a living bog, something which makes it rare in itself. There are many places in Britain which were for thousands of years living sphagnum bog, and which are still very beautiful and fairly squelchy, but which are not sphagnum bog any more. Indeed, that is the case with the majority of places people think of as bogs.

There are fourteen different sorts of sphagnum moss to be found in Britain. Ten of them are bog-forming, and four are more tolerant of enriched conditions. None of the bog-forming sphagna can stand living in running water or in nutrient-rich places, and they all require to be regularly drenched by rain, (though they will survive ordinary British drought conditions,) and to inhabit a soggy environment. Given these basic requirements, they all have their different niches in the bog world, and in different bogs are to be found in different degrees of density. They are not nearly as well understood as botanists would like. Often it is a mystery why such and such a sphagnum is predominant on a particular site, and they all live in various associations with other, vascular, bog plants. It is the presence or absence of particular vasculars which is the best sign of what is happening to a bog and what most threatens it.

On Kilhern Moss, the sphagnum most

Above. Sphagnum mosses are the hallmark of the growing, classic, healthy, acid bog. They will not tolerate enrichment of any kind.

Below. Sundews are carniverous plants, amongst a handful to be found naturally in Britain, mostly on bogs. Their capacity to trap insects and so gain nutrients is crucial to their survival in lean environments.

in evidence is S. *pulchrum*, which is a mar-malade, golden yellow. But there are also S. *papillosum*, S. *rubellum*, S. *magellanicum*, S. *tenellum*, and S. *cuspidatum*. Each of these likes a slightly different amount of wateriness in its surroundings. Some, like S. *cuspidatum*, need total immersion; others like S. *rubellum*, will grow in large, tall, dryish clumps, called hummocks. But S. *pulchrum* will not grow under water and it does not like to be as relatively dry as a hummock. It likes, so to speak, to have its feet in the water and its head in the air. Its dominance on Kilhern implies that there is a remarkably consistent water-level across the whole of this bog. It happens that this is rare in well-developed, thriving bogs, as we shall see.

There are many elements to the bog story, and the mosses are merely the densest, most crucial part of it. They are at times the least obvious. On Kilhern, as on any bog, you are walking through many species of plant, some of which demonstrate the bog's economic value for farmers. Animals, both wild and domesticated, can get a living from the grasses, shrubs, and flowers which grow on bogs. The presence of various grasses is often an indication of a bog's wetness, and thus its well-being, as is the presence of heather or a heather-like shrub, cross-leaved heath.

The bog enthusiast wants to see the fluffy white heads of the common cotton grass, which loves the wet: he is less keen on deer grass or purple moor grass, both of which prefer drier or better-drained condit-ions. In summer the common cotton grass is very distinct with its bushy white top from which Scottish brides used to weave bridal stockings. In winter, there will still be some cotton grass with white tops. The simplest way to tell its presence is from a distance: an area of cotton grass sends a rusty-chestnut colouring across the bog. As they die back from mid-summer onwards, purple moor grass is straw-coloured while deer-grass is a paler beige. The presence of tall heather (sometimes called ling) is another sign that the bog has been drier than it should be. The conditions which are good for heather are, in general, poor for sphagnum, though even a wet bog will support some heather. Indeed, the presence of a good deal of heather, espe-cially in association with deer grass, is taken as the dividing line between moorland and wetland proper.

At the other extreme, the bog-fancier is delighted to see a good deal of cross-leaved heath, which has pale-pink flowers that are larger than those of bell-heather (itself a close relative of heather proper, from a different family). Cross-leaved heath, with whorls of four leaves up its stem, prefers wetter en-vironments than the other two.

Kilhern Moss, like most British bogs, has been grazed with various degrees of intensity over the years. Someone once put

in a drainage channel, which must have caused some damage, though the channel has now been overgrown. Still, there are pockets of enrichment, probably caused by the fertilizing dung dropped by cattle which had been wintered in the farmyard and given special feed. The sign is a classic one: the presence of clumps of a non-sphagnum moss, *Polytrichum commune*, which has sharp-starred leaves and will thrive in mildly enriched habitats.

Silver Flowe, Kirkudbrightshire

Kilhern is either a very young bog or an odd one. It boasts the full bog flora, but has not formed the classic 'ridge-and-pool' system, in search of which we drove east. We finally came to a small river in the Galloway Forest Park. On the other side of this, and sandwiched on either side by steep, rocky, water-falled valley walls, there is an immense tongue of bog filling a great glacier-scraped U-shaped valley floor. This is the magnificent Silver Flowe (many Scottish bogs are called Flows, or Flowes) running from the Round Loch of the Dungeon to Clatteringshaws Loch. It was a breathlessly beautiful day in late October, with warm sun. The eye began to relish the subtlety of the bogs, especially different types of grass and moss, the former making broad, soft drifts of colour and the latter especially lovely in detail. With Richard Lindsay around it was easy to be confident of not sinking up to our necks: he knows his way in such places, and swears they are not often dangerous.

The magnificent Silver Flowe, Kirkudbrightshire.

Without him, so far from anyone, quite alone in an immense mountain-fringed landscape, I would not have been so sure.

We set off across the river and into the bog, which is even wetter than Kilhern. These are the famous ridges, pools, hollows, and hummocks which have proved so enigmatic to botanists over the years. Even Tansley was taken in by them and enunciated a theory called the hummock/hollow succession which has since proved flawed. It appears that a peatbog is remarkably stable. It is a living thing, as we have seen, but it also seems that if it is left to itself its surface appearance can remain unchanged for millennia. Other things (weather, man, or animal influence, for instance) being equal, it will simply get higher year by year. If one mapped its layout of pools and hummocks often enough they would remain the same, perhaps for hundreds and even thousands of years. Tansley was misled by the occasional phenomenon whereby a peat auger will show that succeeding layers of peat in a vertical line through the ages of accumulation have been made by different species of mosses. This appears to have been caused by relatively wetter or drier periods of climatic variation which cause pools to expand or contract, rather than by some process of alternation between the species in botanical succession.

It looks as though a kind of serendipity creates the hummocks and hollows of a mature bog. They form because there is nowhere for water to go; the bog is often brim-full and can store no more. Below it there is saturated peat, and beneath that, very often, impermeable clay. The water is bound to seek out any small depressions in the bog surface. There are so many different sorts of sphagna,

Silver Flowe. A dramatic site for a splendidly lonely habitat.

all of them capable of living in slightly different aquatic regimes, that these relatively wet hollows or relatively dry hummocks are made by whichever type is most suitable to their degree of wetness. Indeed, after centuries, the hummocks and hollows are wholly made of the appropriate mosses. The ridges on a bog will tend to follow the contour lines of any sloped parts.

In a bog formation, botanical life is making geography. Unlike most vegetable matter, in which there is a seasonal process in which leaves and stems die back each winter, contributing dead matter as fertilizer,

An abandoned shooting lodge in Caithness. The former grandeur of shooting parties in the Highlands is now reduced somewhat. Even so, huge tracts are managed for shooting, and not always with much appreciation of what should best be done ecologically, even from the shooter's point of view.

in the case of mosses the old matter merely contributes height to the bog. Mosses grow in clumps. They seldom reproduce sexually, nearly always vegetatively, therefore they expand where they are. The spores which constitute the moss's reproductive material are only produced when the moss is under stress, for they require too much of its scarce energy supplies. The moss kingdom usually only reproduces sexually on the lifeboat

principle that it is better for a new generation to move away from the endangered spot, however risky its chances. A thriving moss will stay where it is, expanding slowly there.

Sphagnum mosses, like all true mosses, have leaves which are one cell thick, though there are two types of cell to be found on a leaf. There are living, chlorophyll-containing cells, and dead cells which are structured as storage spaces with access pores. Mosses grow

in huddles, like pipe cleaners bundled and placed on end, so that only the topmost leaves receive sunlight, carry on photosynthesis, and release oxygen into the air. The structure of the plant ensures that any water that falls from above will be stored. Below the growing surface of the moss there is only lightless stillness in an acid, compacted, sodden world where the dead moss does not decay properly, there being little or no oxygen. In effect, a peatbog is many feet of partly, but only partly, decomposed bog mosses topped by a growing layer. It is the peat-burner's fire, the tomato-grower's plants in a growbag, or any other disturbance, purposeful or not, which completes the decomposition of peat, producing energy as it does so.

Although sphagnum mosses grow in close association with one another, they are remarkably choosy who their neighbours are. Each type lives hugger-mugger with fellow members of its family, but the different types each like a different sort of habitat. Each type is distinct in looks, too; each has its own variation on the water-holding shapes of its leaves; each has a range of colour possibilities within its own share of the colour spectrum. The bog, *en masse*, can look almost wearisome on a grey winter's day, but to its fans it begins to open up its secretive nature and betray its pleasures. The bog as a whole has a weird beauty and is capable of an almost terrible loneliness. When you get down to look at its components in detail, you find, as well as a rich variety of associated species, a lovely variety and subtle richness in the mosses which make it up.

The raft spider, a wetland specialist, photographed on the RSPB reserve at Arne, Dorset, an area famous for its wet heath.

Because different sorts of sphagna, although they all need to be wet, accommodate different degrees of wetness, a bog has a kind of self-limiting upward growth. The hollows which are formed in some places are made of mosses which can grow under water which, having accumulated nutrient from its surroundings, will be slightly richer than the water contained in the rest of the bog. These mosses grow slowly, but they are almost always wet, so the hollow grows upward consistently. However, the firmer ridges and hummocks which constitute the banks of the hollows cannot grow upward too fast, for if they do they begin to lose contact with the bog's own water-table which keeps them saturated.

In a mature bog there are two distinct kinds of very wet areas and three kinds of drier areas.

There are **pools,** called *dubh lochans* (black little lochs) in Gaelic. They are usually defined as being fifty or more centimetres deep, but can be up to twenty feet in some places. Soft peaty silt forms the bottom of pools, which seem to keep pace with the rest of the bog because they collect detritus which blows across the bog's surface. These are in essence without moss growth, though bog bean and other plants may grow in them. Black darter and other dragon-flies and the metallic red damsel-flies love them, with a dragon-fly patrolling a whole or half a pool very thoroughly. Adult water-beetles in abundance live on the larval form of insects in these pools. There are **hollows** (usually defined as being not more than twenty centimetres deep), whose waters support growths of *Sphagnum cuspidatum* and, less often, *S. auriculatum.*

There are **low ridges** (usually defined as being within five or ten centimetres of the water-table) which are the banks of the pools or hollows. Often, at the water's edge, they are formed of *S. pulchrum* and then, a little higher up, of *S. papillosum* and *S. tenellum.*

At this level of dampness cross-leaved heath will thrive best. There are **high ridges,** which are virtually the surface of the bog and the nearest thing it has to a ground level. These are formed from *S. papillosum* and *S. magellanicum.* At this level, but not below it, heather (or ling) will thrive best. There are **hummocks,** which are mounds of moss – sometimes three feet high, but more usually between one and two feet – commonly formed of *S. rubellum.* There are hummock-forming sphagnum mosses: *S. imbricatum* and *S. fuscum* and the non-sphagnum *racomitrium* and *leucobryum glaucum,* particularly in Ireland.

Identifying sphagnum mosses is made more difficult by the way in which their colours vary according to the time of year and to how much light they are getting. Where they sit in the micro-geography of the bog will help. However, there are some guidelines which are worth giving. The following terms identify different mosses:

S. cuspidatum: shaggy, 'drowned kitten' in Richard Lindsay's phrase, markedly feathery in water and always aquatic; green.

S. auriculatum: dark-green to black, and always aquatic.

S. pulchrum: golden, marmalade-coloured; each leaf separate; with brushed appearance, and on pool margins.

S. papillosum: green or pale ochre; bigger and fatter and jucier than most; a ridge species.

S. tenellum: small and golden-coloured.

S. magellanicum: often red, almost like *S. rubellum,* but far fluffier (can be green if shaded or enriched); ridge species; big, fat and juicy; the red equivalent of *S. papillosum.*

S. rubellum: deep wine-red; small, wiry, tightly packed; can be green if shaded or enriched; low hummock.

S. imbricatum: the colour of ginger biscuit; hummock-former.

S. fuscum: hummock-former; ginger-biscuit version of *S. rubellum.*

The main non-sphagnum hummock-formers are:

racomitrium: livid-green hummocks.

polytricum: almost a conifer green; star-shaped, spiky leaves; like green, spiky, bottle-brush.

leucobryum glaucum: very similar to *racomitrium,* but glaucous, greyish green.

Several of the small shrubs which get a living on bogland are found on Silver Flowe: bog-rosemary and the deliciously scented bog-myrtle are there, and bog-asphodel. Of the rather rare white- and brown-beaked sedge, the white form is found. There are also some examples of British insectivorous plants which, like mosses do not have root systems, since they live in places where there would be precious little nutrient beneath them. However, there are insects by the million, and midges and flies provide food for the carnivorous plants. The sundews trap insects by sticky leaf-glands which release nutrients from the victim for the plant's use. All the British sundews occur on Silver Flowe and other peatbogs. The position in the hummock/hollow systems of various of the sundews is usually predictable: the round-leaved sundew is found on hummocks, the great sundew on pool margins, and the long-leaved sundew in shallow hollows. Another bog plant, the butterwort, varies the trapping strategy of the sundews by having a slippery, rather than a sticky leaf-surface.

In the pools of Silver Flowe and other peatbogs, there are rather rare greater bladderworts, whose root-like leaves have small bladders with which to trap water-insects for their nutrient value. These bladders trail from the leaves and have hair-triggers which, when brushed by a water flea, snap open the bladder's lid to trap the unfortunate creature, then snap shut again. End of flea.

This description of typical bog habitats might leave the impression that bogs are uniform. Actually, bogs are not conformist living organisms at all. Some of them are richly eccentric, and none more so than Claish Moss, another western Scottish bog, about fifty miles north of Silver Flowe.

Claish Moss, Argyll

On 19 August 1745, Charles Edward Louis Philip Casimir Stewart, with a handful of companions, was being rowed up Loch Shiel at whose north-eastern tip, at Glenfinnan, he was to raise his father's standard and start what might have become the rebellion which rid the Scots of the troublesome English for ever. Loch Shiel runs inland from Ardna-murchan. On its southern shore there is a splendidly enigmatic and dramatic bog, which Bonnie Prince Charlie would not have been able to see even if his mind had not been on more immediate matters. It is the extra-ordinary quality of mature bogs that they do not change much. What was at Claish Moss more than 200 years ago might well have been there 2,000 years before that.

One of the reasons why the Prince could not have seen the moss is that from the loch's water level one can only see a gradually sloping poor fen slope. The bog proper has built itself an entirely new vege-tative level about twenty feet higher at its centre than the loch's level. This is part of the lovely oddness of bogs. When we look at an immense forest, however high, we are never confused between which is the forest growth and which the ground on which it is growing, even if the forest floor is soil and humus made mostly from the decayed matter of the forest. But on a peatbog the very place where you are walking and the geographical features of the bulging plain made by a raised bog are composed of living matter.

At Dalelia Pier small boats can be hired for general fun or to cross the loch to the moss (and to visit a very beautiful burial island which stands forlornly in the middle of the loch). A steady trickle of naturalists and tourists comes to look at this eerie place.

Claish Moss's 1,000 hectares are formed almost as much by water in pools and hollows as by ridge or hummock. The site is unique in Britain for the expansive way it displays what in Canada or Finland is the classic bog formation in which ridges and pools arrange themselves around the contours of gentle slopes, as though built as terraces on a hillside farm. The only means of progress across it is a kind of bog-trotting in which you chart out what looks a reasonable route in the maze of ridge and pool, and where you go wrong, make a squelching leap from one relatively firm spot to another – trusting that the landing place *is* as sound as it looks. There is no such thing as walking on such a spot, for the entire terrain is moving at your feet. To test its quality you jump on the spot: if you sink a little, there is a chance the bog is healthy. It is a phenomenal place, and it is hardly surprising that there is evidence that otters come on it and desport themselves. Deer pick their way across it. It is the kind of place where you may see short-eared owls, hen harriers, and even golden eagles.

There are the little wispy-flowered cranberry, and bog-rosemary and hare's-tail cotton grass (distinguishable from common cotton grass by having just one stem and one bush of fluffy white per plant). Among the other flora which Richard Lindsay had taught us to expect to see on mosses there was a peculiarity: Claish sports a great many hummocks of *racomitrium*. These are vivid, almost lime, green, and as smooth as boulders. In the afternoon light, they seemed almost to glow. They would normally be a sign of dryness and disturbance in a moss, but Claish, though not as healthy as one would like, has been growing *racomitrium* hummocks immemorially.

Claish is as beautiful as the other bogs we have talked about, with the added excitement of its eccentricity and dramatic location. It boasts its full luxury of mosses, whose textures and looks become so enticing: one

deeper red sort, clumped tightly together so that delving out a particular plant meant plunging the fingers and hand deep into the mound, would shade into a looser, soggy green affair. Mosses on hummocks can seem like velvet. They are lush and glorious and infinitely varied.

One warm afternoon, when we had yomped our way across many miles of bog surface, slushing our way along, learning the business of distinguishing cross-leaved heath from heather and *rubellum* from *pulchrum*, Lindsay suddenly decided that he would show us the bog-fancier's favoured summertime method of cooling the sweating brow. He plunged forward, aiming his face at a soft, inviting, November-chill, clump of *papillosum*. There, face down in moss, he claimed to be demonstrating one of the lesser-known qualities of his favourite plants. Sphagnum moss, being acid and poor in nutrient, was a favoured sterile dressing in years gone by, and certainly up to the First World War. Moss has always had myriad uses. *Fontinalis antipyreticum*, for instance, was so called because it was built into Finnish houses as a fire break. The sphagna and others all make useful dyes and were favoured as bedding.

Exploiting Peatbog: Caithness and Blar Nam Faoileag

There are nearly 700 square miles of Caithness, the exciting, disturbing county in the far north and west of mainland Britain. It is bounded on two sides by the North Sea and on the third by the sharp horizon of the Sutherland mountains. In the gazetteer, you are likely to find the area described as moorland, and that is what, mostly, it looks like. There is genuine moorland in Caithness – places where raised and blanket bog never succeeded the denuded forest and scrub which our forefathers found over much of the terrain – but there are vast tracts which are now not much more than moorland, and where only a few sheep manage to live, where once bog ruled. The problem now, as foresters, peat-extractors and other would-be improvers all have their eyes on these places, is to decide what should be done about ancient bogs in Caithness. We now certainly have the machinery and chemicals, and believe that we have the know-how, with which to wreak great changes in the countryside. It was not always so. Until the Second World War the great wilderness tracts of the country were either run for their game and looked after by gamekeepers who did some of their work far better than their modern counterparts have time to, or were worked by crofters (or their southern equivalents, small-

Right. A deeply-scarred peat bog. This drainage ditch in the mosses and peat merely dries out the bog and helps it erode. *Below*. Long-leaved sundew. One of the excitements of the bogland scene.

Beautiful moorland in (*left*) Invernesshire, and (*right*) Sutherland. Splendid, but dead from the acid bog point of view. These people are shooting over an essentially dead environment which a return to wetness alone would restore.

holders) with a terrific expectation of hard graft.

On the way to the great bog, Blar Nam Faoileag, in Caithness, we drove up the valley road beside the Clyth Burn, where the road goes on to the strange *soufflés* in rubble known as the Grey Cairns of Camster where early man proved himself capable of elegant dry-stone work and astonishing architectural conception. He must have been either hugely committed to celebrating the dead, or been getting a very fair living from the region to have taken time off from food-getting in order to build these structures, like aggrandized igloos in stone. It is a shallow valley which, like thousands in Scotland, displays the dilemma of modern agriculture very clearly. It is littered with derelict crofts, in each of which generations of families, sometimes two to a building, once lived. Around the farmhouses there is the in-bye land which was

once the most treasured part of the farm.

The crofter's family would usually have had a vast area, much of it bog and some of it moorland, on which to graze their sheep, but their farm cottage was usually sited wherever there was some land with decent soils – or any soil at all, if the farm was mostly on bog. Here the ground, however wet, would be drained as best as may be, and large quantities of lime and seaweed applied to make the heather or cross-leaved heath, the deer and moor grasses, or mosses, give way to more nourishing species. This was ground from which hay might be made, or sheep grazed in the hard winter. It was land which, however hard the crofter worked, could seldom match the ordinary meadow of the softer south, but it was crucial land, nonetheless.

Most of it is now more or less derelict on the abandoned crofts of Scotland. It has often been re-invaded by soft rush, usually

the first of the wet wilderness flora to re-establish itself. Some of the in-bye, left alone, might one day return to fen and marsh; given a few hundred years, some of it might one day return to being proper bog. For the time being it bears the same ruined aspect that the croft buildings have: deserted, uncherished. The majority of crofting families have given up, though actually their lives – with modern light tractors and hand tools and knowledge – could now be much easier and pleasanter. However, they did not have a crystal ball in which to see such developments.

In Caithness there are thousands of acres of peaty bog, already largely spoiled, which looked ripe for the new wave of land-uses which foresters and farmers had invented. The land seemed very suitable for forestry and even, in some cases, for re-seeding as relatively high-grade grassland. Before either of those uses, it sometimes looked sensible to strip-mine the peat which had accumulated over thousands of years. Before we look at whether these uses are really wise for Britain's northern and western wetlands, we ought to look at what has happened to bring most of them near to death. They are, of course, prone to the acid rain (caused by sulphur dioxide falling as dissolved sulphuric acid) which is the modern equivalent of pollution from coal-burning factories which killed off the mosses of the Pennines in the nineteenth century: its effect in Scotland is not yet properly known.

Man's hand has, however, had some very obvious effects. The process of cultivation on what had been soggy bogland was much later coming to Scotland than to the lowlands of the south. The pickings were generally poorer, though in the middle of the eighteenth century one ambitious landowner in the lowlands of the Firth of Forth drained great stretches of his land and offered a rent-holiday to brave smallholders. The serious changes probably began in earnest with the introduction of sheep (and the forcible eviction of peasant farmers) in the eighteenth and nine-teenth centuries, but even then many of the massive boglands of Scotland were mostly left alone or lightly grazed – perhaps sometimes over-grazed – by sheep. There was some draining done by tenants and landlords; the remains of their ditches, which can still often be seen, did their bit toward drying the bogs.

The other great change in the use of these highland wastes was the enormous craze for hunting, shooting, and fishing which overtook the rich, and particularly the in-dustrial *nouveaux riches*. Vast tracts of Scotland came to be managed for grouse and deer, and were run by gamekeepers and bailiffs who managed, on behalf of rich masters, vast tracts of Britain's land for game birds and animals. Gamekeepers laboured under two great misapprehensions about the countryside, and to do them credit, until quite recently there was no one who would have gainsaid them. One was that predator birds and animals must be killed to safeguard their masters' grouse. This is ecological nonsense, since it is seldom predators which control prey-species numbers, but rather the other way round. Their other mistake was to believe that, since grouse like the young juicy shoots of heather and cross-leaved heath, it would be best if they burned back the woody growth of heath-ers every two or three years.

Since then, gamekeepers have been bur-ning their masters' boglands and contributing to turning them into moorland. The burning of great tracts of Scotland on a cyclical basis was bad enough when it was done carefully and under well-regulated circumstances, with plenty of keepers and estate hands keeping the fire within bounds and not letting it eat too deep into the flora. Now there are too few keepers, and often, it must be said, there is too little awareness of the dangers involved: the burning cycle is now usually deeply des-tructive. This is doubly sad from the bog's point of view, since there is evidence coming along now that heathers will put forward

much better shoots and be far less woody in their relatively mature growth if they are left in rather wet environments. In other words, if the peatlands were allowed to stay wet, there would be no need to do the burning that is seriously contributing to drying the bogs even more quickly.

Much destructiveness in natural habi-tats is cumulative. We have sketched in some of the elements which contribute to the

spoiling of bog land. Over-grazing will inter-
fere with their association of plants and lead
to trampling which can create small draining
rivulets in the bog surface. Deliberate drainage, undertaken in the belief that a drier bog
is a more productive one, is destructive.
Burning damages the surface texture of the
bog and exposes the living, protective, damp
cover, leaving the peat below vulnerable. The
difficulty with bogs is that although they may

Hill-burning. The standard practice over huge
tracts of land, it is very seldom done as carefully
as was once possible when labour was plentiful.
Now the fires often burn too long, and scar deep
into the peat. There is now much doubt as to
whether burning the heather most encourages
the growth of new shoots and their prevalence
over old woody material. It is thought now that a
return to wetness would most favour the shoots,
the grouse which feed on them, and the shooter.

be stable and capable of enduring thousands of years of changing weather, they are essentially vulnerable.

Blar Nam Faoileag

Blar Nam Faoileag (in English, 'the Bog of the Seagull', perhaps in reference to the arctic skua which breeds there) shows some of the problems in microcosm. A small stone track winds down beside a stream to a small bridge; after that one pushes on on foot.* Half a mile further on there is a typical bogland croft, built on one of the clumps of rock left behind by the glacier which formed the huge, shallow valley. It is surrounded by old in-bye land which has been heavily re-invaded by rushes. You have to walk a good deal further to get to the part of Blar Nam Faoileag which made the Nature Conservancy Council declare it a Site of Special Scientific Interest and buy part of the place. The walking is very beautiful, the footing quite springy. There are streams which cut great swathes down through the peat, and a few deep drainage channels, now much overgrown. These are, of course, signs that all is not well with the bog; the going is not half squelchy enough.

When we plod on to a pool system, we find that many of the pools and hollows have become connected to one another, and instead of being small dammed water patches, appear to be more like little sluggish streams. On the fringes of the bog the surface is dead, though it takes a bog-wary eye to know it. There is plenty of growth of the small bog shrubs and plants: heather, cross-leaved heath, bog-myrtle. But the wetter species are scarce here, and when you plunge your hand beneath the immediate surface to find the rich, vibrant growths of lush mosses in their strident greens and deep reds or bronzes, in

many places there is only a soggy peat to be found. Much of the fringe of the Blar is only hanging on to bog status by the skin of its teeth. It has been too drained, burned, and grazed to maintain the coherent wet-strength bogs need.

Even this degraded fringe of Blar Nam is positively healthy compared with many of Scotland's bogs, which have come dangerously close to being turned into the poorest moorland. It provides, even in its present state, a crucial service which conservationists are increasingly aware many habitats need. It operates as a buffer zone, protecting the essentially undestroyed Blar Nam at its heart, where the kind of bog growth described for Silver Flowe can be found. It is protecting a hugely dramatic place. In the Blar's heartland you are several miles from any road or, what is more, from any shelter. The day we went, we watched a rain storm gather over the Sutherland hills, a grey sky preparing for a grander display. On a dull afternoon with a listless sun biding its time, only occasionally lightening the colours at our feet, it was cool and quiet: good walking weather.

Richard Lindsay who has worked on the Blar for weeks at a time in the summer, says that three scenes in particular characterize it then. First, storms. On Blar Nam Faoileag they announce themselves from miles away. You can see a summer storm coming on the far horizon, with lightning flashes sparking the huge, unmanageable sky. The cloudy centre of the storm is a vast electric theatre, with heavy black clouds trailing beneath them a pale grey haze – the rainfall – and the sunshine hurling a golden light over the tops of the clouds. Golden clouds on top, dark below and grey haze and lightning flashes: "And you realize pretty soon that you're the tallest thing in a landscape of maybe fifty square miles. Very frightening – and you can hear it coming towards you," says Richard Lindsay. "It's like listening to a rain storm coming towards you across

---

*You need to apply for access permission: The Estate Office, Ulbster Mains, Thurso, Caithness.

the sea. And then it's like standing under a waterfall and the whole landscape disappears until an area, perhaps right around you, is illuminated by a lightning flash." Then it passes by, leaving watery bright sunshine.

The day starts again, with a new sharpness, forming the second strong image. Summer days with humming, buzzing, brilliant, lazy afternoons, hazy, with the pools reflecting the blue skies, and dragon-flies patrolling their territories. Settle yourself down on a bog ridge and you may hear the rustle of dragon-fly wings as half a dozen of them dart around the pool.

Much of the Blar is perfect sphagnum lawn, but with a corrugated surface making ridges and hollows, almost in rows. The heart of the Blar is very wet indeed; a walker would sink ankle-deep into any of the ridges. There are masses of cross-leaved heath and some heather. Because it is on the easterly side of

This is peat-cutting on an old-fashioned, domestic scale. At this level of exploitation, the peat-growth can restore itself over time. There is a well-established regime of routine peat-cutting which could sustain itself for ever.

the land-mass, the Blar has continental characteristics, and some affinity with Scandinavia. There is bearberry, typical of Nordic bogs, with its midsummer crop of pale berries, but there is no cranberry here, or on other Caithness bogs; it does not like the drier climate. There is occasional dwarf birch, a short, twiggy shrub which never grows higher than six inches, and sends out a sprinkling of bright-green serrated-edge leaves.

There are extraordinarily deep pools. The water is clear but brown, and in its dark depths one can see a towering column of bog bean and *cuspidatum* rising from the depths. The bog bean will be in white and pink flower at the surface. Water-beetle and dragon-fly larvae are suspended in the water. There is not enough oxygen for fish, but teal will breed on the pools. However, it is the red-throated divers which most characterize them, with their haunting, cry, a melancholy wail echoing out across the landscape. In drier places, there is bog myrtle, also the lichen, crottle, which families used to wander the bog to find, and use (having steeped it in urine) as a crimson or purple dye.

This (*left*) is the kind of deep swathe that modern equipment can cut into peat, making furrows ready for tree-planting. We have the technology to do all sorts of large-scale operations in fragile environments, but that does not mean that we know how our new operations – forestry for instance, or big scale peat cutting (*below*) – will survive against the greater ecological test: the future.

On a still warm day, when the dragon-flies are out, patrolling their territories, you may see a hen harrier drifting gently across on still, V-spread wings, looking for young grouse or meadow pipits. Hen harriers always travel low, as do all birds on bogs except the rather rare eagles. The merlin, for instance, hurtles across very low down, doing aerobatics, looking to pick up meadow pipits and skylarks. There are also the horse-flies, about the size of a house-fly, but, at rest, with wings lying back along the dull brown body, and with two big rainbow-coloured, multi-faceted eyes. They are silent in flight except when coming in to land, when there is a low-pitched buzzing. Their victims feel nothing until a sharp stab as the sucking begins, and the fly sits there, seeming suddenly very malevolent, sullenly intent on drinking blood. "They won't shove off, even when you strike out at them," says Richard Lindsay. "You tend to swell up spectacularly, and very painfully. Clegs, peat-cutters call them: hateful. They don't stop you working, but they are irritating and the swellings hurt later."

The Blar has other denizens which are not always benign. From early spring there are arctic skuas, whose chicks hatch in May and fly by June. These skuas, seagull-sized, have the habit of dive-bombing intruders, which can be alarming. However, it is the mosquitoes which stop you in your tracks. These are *ceratopogonidae*, biting midges. They are liable to be out on a summer's day or in the evening if it is warm or still. Attracted to the warmth and the carbon dioxide exhalations of the people inside a tent, they will often greet campers first thing in the morning, coating the outer tent black with their bodies. They cloud rather than swarm, making the whole landscape shimmer. When they are around – usually on Blar Nam for at least a few hours during any summer week – one cannot breathe freely because each intake sucks ten, twenty, thirty of the creatures. "You can't open your eyes, and you're

half-crying. You can't see, can't breathe, and you're being chewed to death. And then you come up in lovely little lumps the size of a half-penny coin. You moan a lot, scratch a lot. You end up fully dressed in all your waterproofs, sweating madly. Sometimes we've even had to abandon a day's work and retreat inside the tent. Usually, you just carry on, or try to find a place where there's some breeze."

Come the dusk – and if you are spared the mosquitoes – the Blar takes on a quieter, stiller mood. The short-eared owl is more likely to be seen now – flying on stiff, bowed wings, drifting by absolutely silent. Dunlin

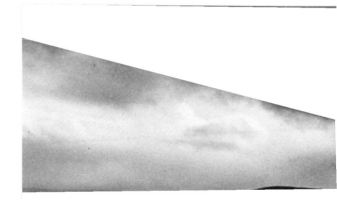

and golden plover both breed on heather. The dunlin is perhaps the typical bog bird, laying eggs on top of hummocks after wintering on estuaries. It is usually associated with golden plover; on the Pennines, dunlin are called the plover's page. Plovers have a kind of weeping sound, while the dunlin cheeps and trills, especially as it tries to alarm an enemy with its characteristic sudden tumble toward the ground, pulling out of the dive at the last moment, crying as it goes.

In the quiet of the long Highland dusks you can read a book till well after midnight, and it is properly dark only till three or four in the morning. In the Blar Nam Faoileag midsummer midnight an eerie loveliness comes upon the place – two or three sheep bleating to one another, an occasional bird call, maybe a dunlin trill. If you have camped near a ruined croft with the in-bye land going rushy there will often be snipe, beating the bounds of their territory, and you may hear the wonderful sound of a snipe 'drumming' in another controlled bird freefall in which two erect feathers produce a thrumming sound.

Eroded peat bog in Invernesshire. No-one knows the long-term effects of wasting our peatlands. It may be that within a hundred years most of the peat in this picture will have flowed down to the sea and into reservoirs.

We walked long hours on the Blar: it is a place where one easily feels lost, and where it would be easy to become really lost. The horizon is immensely wide, as are the horizons in almost all wetlands. You can get no perspective in such a place, and no vantage point. You cannot climb some small incline to get the lie of the land, for there are no inclines. You might walk within a hundred or so yards of a major feature – say the lovely Loch Ruard – and not see it. You can walk within two or three feet of one of the old drainage channels and miss it. You can become disoriented in a bog-scape as easily as in impenetrable woods. True, there is, on the Blar, always the great line of mountains to the south west and west; the long, bumpy outcrop of Scaraben, and the perfect, breast-shaped dome of Morven. But on the kind of day we were there, in early November, a mist might descend at any time, and blanket out these distant landmarks.

It is the way of things that the beginning and end of a day tend to be the loveliest parts. It was certainly true that day. In the mid-afternoon we had watched heavy rain clouds swathed in mists march toward us, making the pale sunlight shaft down on the emptiness of Caithness. We were spared the downpour, but later, as the chill increased, there was a fine spit of misty rain while we trudged along. Over on the mountain horizon the sun was going down (a combination of winter's low-flying sun and the presence of mountains made the Caithness sunset occur at around half past four in the afternoon) and a great fiery cyclorama of hot pink was thrown up behind the sullen mountain shapes. As evening drew on we trekked back toward the abandoned croft, back through the less-than-perfect parts of the bog. Everywhere was still beautiful. It is understanding these places better which makes for a view of them both more appreciative of their loveliness and more anxious for their preservation.

Glasson Moss

We had become mightily fond of the big peat-lands of Scotland, and increasingly upset about the decline of many of them. To find out about bog recovery we went down into England, but only just. South of the Solway Firth there is a wet, low piece of Cumbria which is littered with boggy names. Among them is Glasson Moss, near Bowness, which, miraculously and probably only because it was *very* wet, was never reclaimed for agriculture. Part of it was used for commercial peat exploitation, and that part is going to take many years to recover.

One site of immediate interest is a spot where during the drought in the summer of 1976, there was a devastating fire which burned tens of acres to a crisp. In many places the fire went down to bare peat. It was, in a sense, just a typical example of what can happen, and is happening, on all of the British boglands. But it happened on a National Nature Reserve, and the Nature Conservancy Council therefore felt strongly committed to seeing what could be done to restore it. Glasson Moss has become a test-bed of moss reclamation: more particularly, it is a test-bed for the reclamation of damaged bog as rich bog.

In November, when we visited it, there was little enough sign of vigour on the place, with none of the bog flowers or shrubs in flower, but Richard Lindsay was vastly encouraged. There were plenty of signs that the mosses were recovering. A two- or three-year-old footprint is often clearly in evidence on a bog surface, which makes one realize how impressionable bogs are. But they can recover.

A great sundew. Exactly the kind of lovely (carnivorous) plant which will increasingly become rare as we lose wet, healthy peatbog.

CHAPTER EIGHT

Estuaries and Saltmarshes

Previous chapters of this book have dealt with very wet-footed places. Some of them, especially freshwater marshes, are very variable in their wetness, perhaps even dry for months and only wet for a few days at a time. The salt-water marshes of Britain add to ordinary variability two extra problems. Across the extent of any estuarine marsh there is the factor of tides and the peculiar daily alternation between salt and fresh water.

Saltmarshes are formed of a constant accretion of silt brought from inland by the flow of fresh-water rivers and deposited in more or less sheltered valleys or bays, but they are strongly influenced by the twice-daily influx of salt water. Between them, these factors distinguish the estuarine range of habitats – almost dry and almost fresh-water grass meadow at one extreme, perpetually wet and very salty mud-flat at the other. They add up to a wonderful treasure for wildlife, which has been preserved for the present generation mostly by the physical impossibility or great expense involved in 'improving' it for agriculture. Even so, almost everywhere parts of it are threatened by the power of modern man's technical ability, and in places by his carelessness in controlling it.

At the wet-and-salty end of the spectrum, down in the mud-flats where the sea's influence is strongest, very few plants can get a living. The water is often too salty, but more than that, the incoming tide stirs up the mud's surface so drastically that very few plants can get a grip. Sir Arthur Tansley, to whom anyone trying to refine his appreciation of the British habitat owes such an immense debt, drew a distinction between those mud-flats where only green algae and sea grass (sometimes called sea wrack, *Zostera*,) grow, and those which he calls proper saltmarshes, which are characterized by glasswort (or marsh samphire). This plant colonizes the mud which has been partially stabilized by the algae, but it can only get a firm grip in flats which the tide leaves in peace for at least a few days in succession.

Some of the Estuaries and Salt Marshes of Britain. This sort of habitat probably represents the largest single resource of wildness in the country. There are tens of thousands of hectares of estuarine habitat where the public has free access and where wildlife is rich. The map and chapter can only at best sketch in a widely-scattered sample.

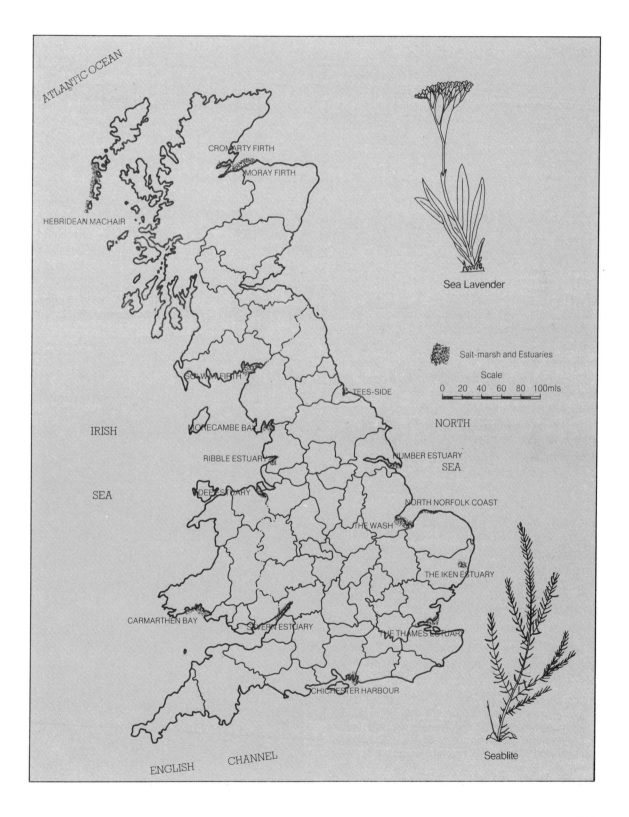

ATLANTIC OCEAN

CROMARTY FIRTH

MORAY FIRTH

HEBRIDEAN MACHAIR

Sea Lavender

Salt-marsh and Estuaries

Scale

0   20   40   60   80   100mls

SOLWAY FIRTH

TEES-SIDE

IRISH

NORTH

MORECAMBE BAY

RIBBLE ESTUARY

HUMBER ESTUARY

SEA

SEA

DEE ESTUARY

NORTH NORFOLK COAST

THE WASH

THE IKEN ESTUARY

CARMARTHEN BAY

SEVERN ESTUARY

THE THAMES ESTUARY

CHICHESTER HARBOUR

Seablite

ENGLISH   CHANNEL

Certainly, in the sea-grass-dominated areas, only wading birds, feeding on the huge numbers of invertebrates which live in the mud, or stranded yachtsmen stand on them at low-water, and the latter only briefly. Mud-flats are rich winter feeding grounds for curlew, bar-tailed godwit, oyster-catcher, redshank, knot, and ringed plover, each with its own purpose-designed beak for probing the mud for such creatures as *hydrobia*, lugworms, and ragworms. These species achieve enormous numbers in estuarine mud: 60,000 *hydrobia* a square metre have occurred. On the other hand, birds consume them voraciously: 3,000 snails have been found in a single bird's stomach. Mud-flats are a permanent home for birds such as shelduck, while for others, especially certain ducks, they are temporary feeding grounds. For the wading birds and wildfowl wintering here, they constitute a vital feeding ground. The mud-flats, especially, are crucial to them when the tide allows, while the high saltmarshes and nearby, often reclaimed, fresh-water marshlands and farmlands serve as roosting and high-tide feeding places.

The Solway Firth, one of the country's loveliest estuaries, is a classic mix of mudflat and 'upper' marshes where cattle are grazed. Various attempts were made to turn it into a major port during the nineteenth century, but they failed: of which we can probably be glad, since many other estuaries have borne their fair share of industrial depredation, whilst this one at least remains benignly agricultural or even completely wild.

I first learned about estuaries by trying to sail small dinghies on those alarmingly, indeterminately, shallow waters. This was at Itchenor and Bosham in Sussex, dinky little sailing resorts both, where weekend salts rightly sought out, and seek still, the world they remember from books like Arthur Ransome's *We Didn't Mean to Go to Sea* (though it was years until I was stuck on the mudflats at Bradwell, the Essex equivalent, where that book was set). Its charm for me then, and I think now, was that to sail or walk in such places was to be a part of the sea's

world but not to essay the full impertinence of finding one's recreation on the sea itself. An estuary has its dangers, especially when tide races come in across the flat-scape quicker than a man can run. They can be very big and splendid, but one usually remembers not high drama but a lovely combination of land and water, the one easing into the other before the eye, encouraging relaxation and reflection and a certain confidence.

There are few proper saltmarshes in this part of Chichester Haven, though there are some at Wittering, by the fragile and

splendid sand-dunes which form on that beautiful spit. But there are mud-flats; in Langstone and Chichester harbours combined, there are something like 3,200 hectares of mud at low tide. This is the wettest of the estuarine 'terrains'. It is covered by water for less than half the day, but for a part of every day, winter or summer. As Sir Arthur Tansley, charting the 'zone' changes marking the succession of one habitat to another, notes, proper saltmarsh communities are free of water for much of the day, especially in summer.

The mud-flats, stretching grey and treacherous – especially as viewed by a yachtsman who has got stuck on them, left low and soggy by the retreating tide – are still, infuriating places. At their worst, they can be leaden or gale-lashed and desperately lonely. When I knew them most intimately, from their watery fringe, I was an almost criminally incurious teenager for whom they were no more than a backdrop for boating.

Their apparent barrenness should not deceive us. The mud-flats may look improbable food banks, but the lugworms, ragworms, and snails, surfacing to feed on the beds of algae which colonize the mud surface and on the organic detritus left behind by the retreating tide, make estuaries among the world's richest converters of sun into food energy, though much of their richness is sent down to the sea. Some of it leaves under water, in the bellies of fish. Some of it, of course, is flown out by birds. At Chichester and Langstone harbours, extensive mud-flats and beds of sea grass, or sea wrack, help support 6,000 dark-bellied Brent geese, and the invertebrates around 6,000 shelduck. There are huge populations of waders in winter, upwards of 50,000 birds: dunlin, redshank, curlew, bar-tailed godwit, and plover. The birds are part of what makes the estuarine stretches of Britain's coast so lovely, but even if there was not the cry of birds to keep you going, these would always be places to walk on with enthusiasm.

The wildness of the Solway scene indicates why both birds and people in search of solitude like such places.

However, at 3,000-odd hectares, the Chichester/Langstone mud-flats and small saltmarshes are on an almost suburban scale of wilderness, compared with the 12,500 hectares of the Dee estuary, 16,000 on the Severn, nearly 13,000 on the Humber, 6,000 on the Solway, 17,000 at Morecambe Bay, 20,000 at Carmarthen Bay, or 8,000 on the Ribble estuary. These are vast places, each with its own vast bird numbers – 125,000 waders on the Dee in winter, well over 250,000 on Morecambe Bay, and 95,000 on the Ribble, for instance.

Anyone who has sat on the train which runs down the South Wales coast, skirting Carmarthen Bay, will remember the silver-grey, glinting, smoothness, of the mud-flats there, especially with a low, dim, winter sun glancing across it. It boasts a huge population of the common scoter. Anyone driving on the M4 across the Severn Bridge should be tempted to motor down to Sudbrook, where there is a little car park, and access to the mud-flats, and a sudden tranquility after the streamlined, roaring world of the road.

The Wash, with 23,700 hectares, is the biggest mud-flat in the country, and supports major geese populations – especially pink-footed geese and dark-bellied Brent geese (feeding on algae) – and duck. Something like 280,000 waders, particularly dunlin and knot, winter on the Wash. This is a phenomenal place, a place to visit in the full blast of winter, perhaps, when your arrival will be along deserted roads in a deserted countryside. You carry on, huddled in coats and scarves, and come very suddenly to the sea-wall, like a castle's parapet. Up and over, into the full force of the wind, and the whole immense scene is laid treacherously before you. There are a few grey seal here, but up to 5,000 common seal, probably the biggest population in Europe.

However, the Wash has more recently been even more famous as the scene of one of the few occasions when the conservation lobby took on the Ministry of Agriculture, Fisheries and Food, and won. Gedney Drove End, near Holbeach in Lincolnshire, is about half way along the King's Lynn/Boston line of the Wash's inland edge. Gedney Drove was an old cattle track, leading down to the summer grazing which was the best that could be managed on the saltmarshes. Since improvement, the land inland of the sea-wall had become highly productive, with bulb fields and winter barley as profitable crops on what had once been grazing land. In 1980 there was a public inquiry as to whether further parts of the saltmarsh to the seaward of the wall ought to be 'improved' – that is, defended from the sea, drained, and claimed for agriculture – and, some of it, sold off to prospective customers at £2,000 an acre. The case was lost, though on something of a technicality in planning law; no principle appears to have been established that such marshes are more valuable for wildlife than for agriculture. In this place the watery fringe between land and sea, mud-flats and salt-marsh, remains for the birds.

Saltmarsh is defined by the regularity and duration of its flooding or (and this amounts to the result of the first characteristics) the kinds of plants which grow there. Thus, the first of the true saltmarsh plants define the transition from a land which belongs mostly to the sea, to land which is often almost cricket-pitch-firm and flat and lawned. Importantly, the saltmarsh communities vary, in quite definite zones, between those that are covered by sea water for half their lives, through to those that are covered by sea water for only a few hours in the spring and autumn high tides. The most vigorous saltmarsh plants do not thrive merely by colonizing the mud and somehow managing to hang on; they are constructive plants which combine the capacities to shed the tidal mud smoothly from their upper stems, thus being able to photosynthesize, and to trap silt lower down. This in turn helps the

saltmarsh gradually to stabilize.

The first of the saltmarsh plants to colonize the higher mud-flats is the glasswort, also called marsh samphire. As young plants colonizing the mud (which will often have been somewhat stabilized already by algae), they cannot withstand the action of the tides, and therefore need a few days' respite from tidal action to allow a proper community to develop. Glasswort can readily absorb by osmosis whatever fresh water is around it; it has a strongly salty sap. Sea aster and sea manna grass (sometimes called common salt-marsh grass) may grow alongside the glass-wort, and will be more and more the co-dominant species as the 'lower' marsh gives way to drier marsh.

In 1870 one of the parents of a very vigorous competitor to glasswort 'escaped' from a ship in Southampton Water. Rice grass (sometimes called cord grass) is a hybrid of

Hulks of barges at Chetney in the Thames Estuary.

European and American species, and it transformed parts of Poole Harbour very quickly. Between 1911 and 1924, as two extraordinary pictures in Tansley's *Britain's Green Mantle* reveal, soft mud-flats there were transformed into densely-growing saltmarsh. Rice grass, now found as far north as Holy Island, the Mersey, and the Outer Hebrides, is sometimes planted as a deliberately aggressive colonizer of high mud-flats or unstable saltmarsh. It operates as a powerful binding force, growing vigorously and trapping moving silt so that it can begin to form salty, permanent soil.

Other saltmarsh plants, especially sea purslane, sea aster, sea lavender, and sea thrift, flourish almost on a desert-plant principle. They often have extensive root formations, fleshy leaves, or a hard cuticle. Some of them also have very pretty flowers. Sea lavender makes lavender-purple blossom in middle and late summer, while the drier-preferring thrift (sometimes called sea pink) has rose-pink flowers in April and, especially, May and June. Sea purslane is a sage-grey shrubby plant which colonizes especially the upper banks of steep creeks, appearing to like the rapid drainage of such sites. At the Isle of Sheppey, in the crook of the elbow formed by the curving Shellness spit, there is saltmarsh of very extensive sea purslane, probably because this is a place with myriad deep creeks in soft mud. These 'middle marsh' communities can often be grazed. Sheep have grazed the Dovey Marshes of the Dyfi estuary for centuries (though in the lower marsh, rice grass is making big inroads on the glass-wort community).

As the silt accumulates in the lower marsh and it becomes more established and, most important, less frequently inundated, other plants join the sea manna grass and the sea purslane as powerful influences. Occasionally, at places such as Blakeney Point in Norfolk, the rare sea heath, with rose-coloured flowers, comes in. Sometimes a maritime form of the red fescue grass dominates (especially in sandy silt), and sometimes, in terrain which is less frequently inundated and less salty, stands of sea rush begin to form. This higher marsh is very often heavily dominated by common saltmarsh grass and red fescue; large areas of the lovely salt-marshes of the Solway are of this sort. Grazing on this marshland produces a smooth, close-cropped, lawn-like texture which is splendid to walk on. Called merse, it provides food for three-quarters of the wintering population of Spitsbergen barnacle geese.

On the south side of the Solway Firth,

Solway Firth. The tolerance of plants to different conditions creates sharp divides between channels, and lower, and upper marshes. The distinctions tend to be self-perpetuating and rather stable.

Overleaf. The saltmarshes beside Shellness on the Isle of Sheppey in the Thames Estuary provide a fine marshland habitat very close to industry and the huge populations of London, Essex and Kent.

in the Solway Coast Area of Outstanding Natural Beauty, there are some beautiful stretches of this terrain, deeply cut by the creeks which wind across the billiard-table surface. The channels in a saltmarsh are at constant odds with the flat-topped, raised hummocks formed by the grass communities.

The scour of the tide, especially at ebb, is constantly deepening the channels and trying to erode the hummocks. On the other hand, the scouring action of the tide in the channels creates the kind of aerated soil favoured by strongly binding plants like sea purslane. The tides and the hummocks are usually well-

matched competitors for space, and create a stable environment from their silty world.

Though these great estuaries provide huge vistas, the estuarine habitats of the British Isles are not all grand or extensive. At Orford, in Suffolk, for instance, there is the curious excitement of watching sailing-boats

biffing along against the breeze on the water which is almost overflowing on one side of the path, while on the other, three or four metres down the steep slope, the drained marsh stretches away. Meanwhile, at walk's end, at the Orford Oysterage, there is the best lunch in England, proof that muddy Essex and Suffolk waters still bring forth excellent examples of what were once part of the staple diet of the British, though even so marvellous a place cannot make the fare as cheap as it must have been once, when people begged not to be paid in kind, in oysters or salmon.

But then, the estuaries of Britain do not carry quite the oysters and salmon they used to. Pollution has had a good deal to do with the decline. We have fairly good river-pollution legislation, but industry has persuaded governments not to implement many of its provisions. Pollution of the sea itself is likely to prove even more intractable. There

Solway Firth still earns its living for man as well as wildlife. Its merse is used for cattle grazing, and the luck of the place is that it has never been the subject of a thorough reclamation scheme. The pickings can be rich for people who can get permission and funding for reclamation.

is still cockle-farming in the Thames estuary, but it used to be a famous place for oysterages as well. The Thames is different now. Much of the traffic on the lower, estuarine reaches, is devoted to a downward flow of rubbish barges (which used once to return laden with straw and hay for London's horses), and to the vast tankers which, viewed from across the remaining marshes, perhaps at Chetney, seem to stalk their way through the countryside. The green meadows and arable land lap them, while the low hills of the Isle of Sheppey rise behind. Across Chetney Marshes, approached from Strawberry Hill, there are some fine little footpaths running beside the Funton Creek and the little channel, the Shade, which runs down to Slaughterhouse Point and thence to the River Medway.

At Bedlams Bottom, on the shore of Funton Creek, there is a sad, rather beautiful graveyard for old wooden barge hulls. On a searingly bright afternoon in February, this is a wonderful place to see: a pretty, muddy, estuarine place, with a low hill to one side, and to the other a small sea-wall with sheep grazing on reclaimed and drained terrain which still shows the remnants of the channels of the saltmarsh this place used to be before improvement. A walk by the Funton is all the more extraordinary because it is so near London. These are precious areas of wilderness so near to big populations. To match their beauty, you would have to go to classic places: to Blythburgh, where a commanding, high-windowed church stands at the head of a lovely estuary inlet, and Angel Marshes on the River Blyth, or to the footpath by Troublesome Reach on the River Alde, at Iken. Either of them would do, but perhaps the latter is especially evocative with its strong reminder that this is Benjamin Britten country.

The range of threats to estuarine and coastal wetlands is wide. On the unusual wet machair sites of the west coast of the Outer Hebrides, where shell sand has formed the basis for grazing lands, often with lochs and shallow pools, valuable bird life is threatened by EEC-sponsored agricultural 'improvement'. The loss will be apparent to anyone who knows the islands. From the experience of a wind-blown morning we spent at the machair at Northton in the south of Harris, when sand-grains and sea and rain were all ferociously blasting along horizontally as we struggled to see the ruins of a chapel, we understand how easy it is to love such places. On the Cromarty Firth (where up to 17,000 wildfowl and around 9,000 waders overwinter) there has been a long-running threat of industrial development. The Severn estuary is the subject of constant debate as to whether it could or should become Britain's first tidal power site.

It has become very important now to preserve estuarine habitat wherever it is threatened. Already the inter-tidal mud-flats of the Tees estuary have been reduced from around 2,400 hectares to less than 175 in 125 years. Southampton Water, the Tyne, and the Mersey have all seen huge developments on their estuaries. We cannot afford to allow the process to go on.

Mudflats on the Solway Firth. The mudflats of estuaries may look bleak and barren, but they are very valuable for the invertebrate life which teems beneath and on their surface, and for the algae which colonise them. These are crucial to birdlife, especially visiting waterfowl and waders.

CHAPTER NINE

The Conservation Case

One Wednesday in February 1983 I was telephoned and told confidentially that Sir Ralph Verney had announced to the council of the Nature Conservancy Council, of which he was chairman, that he had been sacked. Would I make sure that the truth of the case was known by newspapers and television? Sir Ralph had been sacked by a newly appointed Secretary of State for the Environment; he might be prepared to say that he had resigned, but it was important, my informant stressed, that the real facts be known.

Next day, *The Times* diary ran a story that the Environment Secretary had been pressured into sacking Sir Ralph by two Tory MPs who sit in the Commons for constituencies near the Somerset Levels, part of which Sir Ralph had recently agreed, finally, to designate as a Site of Special Scientific Interest. Since it is the NCC's statutory duty so to designate any site which meets its criteria, and since it had been proved that the place involved, West Sedgemoor, did meet the criteria, the position seemed to be that a Government Minister had sacked an official body's chairman for doing his job.

I tell this story because it helps illustrate the peculiar absurdity surrounding the conservation issue in the country. Sir Ralph was no young hothead: he had, rather, a reputation both as a wealthy landowner and a patrician, of being rather inclined to do the NCC's business behind closed doors, dealing with his fellow landowners on the basis of personal friendship rather than through officials. Curiously, he had fallen victim to a bizarre situation in which the farmers on our more valuable wildlife sites cannot lose – whether they conserve or wreck the countryside. The difficulty is that the farmers do not realize it. They still behave as though having their land designated an SSSI was going to cost them money. They seem to see it as a kind of penalty instead of an accolade. Perhaps they resent it because though many of them are fond of saying that farming is the best guardian of conservation values, all too often it has been proved conservation's worst enemy. However, under the Wildlife and Countryside Act, once their land has been designated an SSSI, farmers can either negotiate compensation to keep the land as it has always been, or, after a statutory period of negotiation, go ahead and 'improve' it.

Draining our wet fields and marshes and bogs has pressed much land into service which would otherwise have been of use only to wildlife. Many of the schemes have not worked or been profitable. Many have soaked up huge quantities of public finance and benefitted very few people. Nonetheless, many schemes are worth doing, especially when they return to good production old in-bye land which has become derelict, as these men are doing in Dumfriesshire.

In a crazy system, two different Government departments are offering enormous sums of tax-payers' money to achieve precisely opposing objectives. The Ministry of Agriculture, Fisheries and Food (MAFF) is spending a fortune tempting farmers to spoil the countryside in pursuit of quite phoney 'productivity', while the Department of the Environment is proposing to spend even more money on bribing farmers to resist these blandishments. The problem is the result of a very long-standing misreading of the farming practices of the country. For too long farmers have persuaded the public that Britain must produce more and more milk and meat, while the conservation movement has been too slow in pointing out the dangerous and expensive consequences of much modern farming practice. A fiction has arisen by which farmers have sounded sensible and reasonable when they have said that if the nation wants conservation, then it must pay for it.

Critics ranging from the Conservative MP Richard Body (with his recent book *Agriculture: the Triumph and the Shame*) to Marian Shoard (*The Theft of the Countryside*) have castigated the maze of grants and subsidies which bolster farmers' incomes. The Ministry of Agriculture, has for years been grant-aiding 'high-input, high-output' farming. Left to themselves many farmers would continue to farm as their predecessors did, but MAFF, in spite of a statutory obligation to take account of conservation, has often tempted them with grants to drain or plough land in order to increase production. It also increases their dependence on fertilizer, pesticides, bought-in seed, diesel fuel, and machinery, much of it also subsidized. Cynics in MAFF admit that we have no need of such production, most of which is in surplus, but insist that every pint of milk and bag of grain helps Britain gain EEC support funds. Until recently, MAFF often steamrollered this profligate and pampered agriculture through

opposition from the Department of the Environment.

Farmers, uniquely in the business community, do not face serious planning control as they alter the face of the British landscape. More than that, the Wildlife and Countryside Act proposed a new and expensive system of compensation for farmers who want to change their farming practices and who might ordinarily expect MAFF grant aid for the capital expenditure involved, but who are restricted on conservation grounds. "It is proposed that farmers be compensated for the profit they would have made," says John Bowers, an economist at Leeds University, "but, typically, between fifty and eighty per cent of that profit would have been subsidy, and sometimes more." No comparable principle of compensation exists elsewhere in the economy.

"It does seem peculiar," says Richard Body. "After all, a farmer buys a farm knowing that it has, say, wetland on it, and gets it on the cheap because of its lower productivity. It appears rather naughty to expect to be paid handsomely for forgoing subsidy from the tax-payer to improve the value of the land."

Before his sacking, NCC chairman Sir Ralph Verney said in 1982 that he would need £20 million across ten years to buy conservation agreements with farmers who eschewed MAFF grants. However, the NCC needed £2 million in 1983 alone for this purpose, and this before the flood of deals under the new scheme begins.

We are talking about big sums of money. One farmer in Kent had some potentially valuable wet fields he wanted to drain. He was offered £100 an acre (about 0.4 hectare) per year to compensate him for leaving them as they were – profitable, handsome and useful to wildlife. It added up to £100,000 a year. In other words the farmer stood to have been made a 'conservation millionaire' as well within a decade. Richard Body's book showed that the consumer and the tax-payer are between them shelling out about £13,000 per year per farmer in Britain in subsidies. Some of these subsidies involve paying far more than the world price for food commodities, some of them are tax and rates privileges, while others are grant incentives to 'improvement' work. However, many small farmers receive hardly any of this support: others, the big farmers, receive a very high proportion of it. Modern agriculture support is a system in which the rich get richer and the land gets poorer.

What is the scale of damage?

First we must stress that in the great generality of the landscape, in the soggy meadows and riverside fringes, we have seen huge yet piecemeal devastation. It is a process of attrition and diminution which has gone on unobserved for centuries and continues apace even

now when it is urgent to call a halt.

There is value for wildlife in almost any site where old farming practices have been preserved, either by the whim or dedication of a conservation-minded farmer, or where the land is marginal and until recently not worth improving. In England, Scotland, and Wales there is a total of about a million

This is the River Rea in Birmingham. It has been wrecked in many places like this where people used to have a walking spot amongst trees and shrubs and flowers. A campaign has been mounted to make the rest of the drainage scheme less destructive: there was no need for it to have been so damaging.

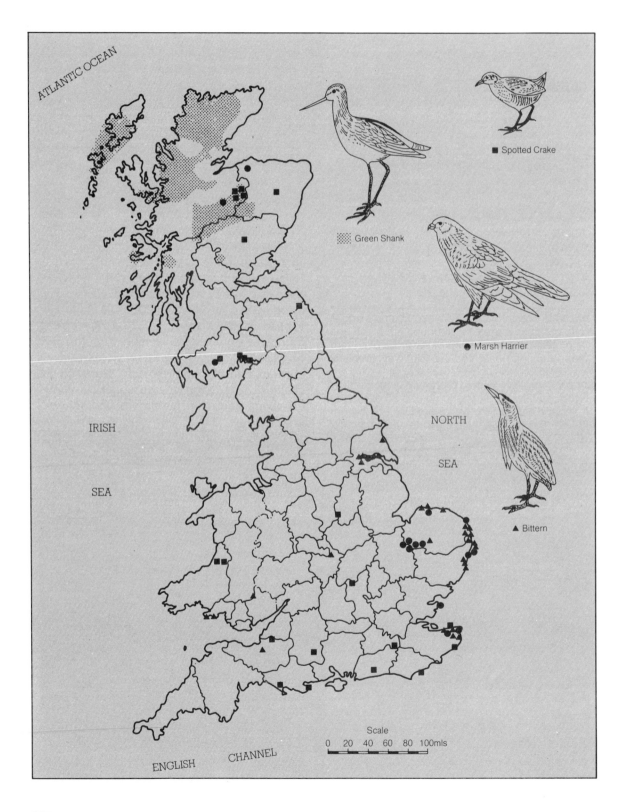

ATLANTIC OCEAN

Spotted Crake

Green Shank

Marsh Harrier

IRISH

SEA

NORTH

SEA

Bittern

Scale

0 20 40 60 80 100mls

ENGLISH   CHANNEL

and half hectares which have been designated Sites of Special Scientific Interest by the Nature Conservancy Council. These include much land, whole mountains in places, where there is no possibility of competition with agriculture. There must be thousand upon thousand of hectares, expecially in piecemeal lots, where wildlife can flourish, but which have not come to the notice of the NCC or are too small to seem worth designating. These will include long stretches of lovely, scrappy, riverside fringe and thousands of farm ponds. (A Ministry campaign to grant-aid filling these in produced such an outcry that the policy was scrapped, and volunteers offered to dig them out again.) They take in myriad soggy patches in poor grassland. They are what make the country worth walking in, but are worth all too little to the modern farmer.

If we look at the most grand and primitive wetland scene, the vast acid bogs, the devastation has been horribly dramatic. Hill and lowland peat sites have covered the British land surface with around 1,582,000 hectares of peat, around seven per cent of the land surface of the UK – the same, very roughly as is covered by urban development, or by SSSIs. However, much of this overlaps considerably with the twenty-one million hectares of the UK's land surface which is devoted to agriculture and forestry (eighty-six per cent). Much of our peatland has been swallowed up by the agricultural maw. Almost all the lowland peat was – arguably quite sensibly – destroyed many hundreds of years ago.

Richard Lindsay of the NCC says that he and the peatland team of the Nature Conservancy Council have walked perhaps 16,000 kilometres in Scotland in search of the Scottish bogland. Ninety per cent of what they saw was damaged. On lowland raised bogs it has been shown from a survey of the history of 120 known sites in the 1840 Ordnance Survey map, that many sites entered the twentieth century already partly 'reclaimed' or damaged. Peat-cutting, forestry, and especially agriculture had all taken their toll.

Below. Pollarded willows are a very rich wildlife habitat. Their boles provide flowers, insects, birds and mammals with niches. It costs about £12 to pollard a willow, which will keep it growing safely and healthily for years.

Some of the prime wetland birds of Britain. These are just a very few of the birds which depend on a healthy, extensive and relatively unfrequented wetland habitat in the country. The less such habitat occurs naturally, the more there will be a need for conservation bodies to establish reserves of their own: if agriculture is encouraged to wreck the wetlands with tax-payer's money, then conservation bodies will have to protect them using private subscriptions, since it is unlikely that public funds will do the trick alone.

Above. Deep drains in old wetland will often be polluted by chemical reactions in the soil. This is 'ochre' on a Somerset level: few people anticipated the difficulty. *Right*. How a digger-driver does his work, and whether or not it is done with sympathy for wildlife, is often in his own hands, and could be made a matter of pride in an extension of his skill, especially if the piece-work rate were adjusted in favour of habitat preservation.

By 1950, heavily-subsidized conifer forestry had taken a huge amount, with agriculture hardly taking any more in the past hundred years. By 1978 forestry was claiming more and more, with agriculture having claimed in the interim about the same number of hectares that it had in 1890. In about a hundred years, forestry had taken half of the 12,000 hectares of the study sites.

Now our too-dry bogs are falling prey to the destruction process. From the blanket bog of Robinson's Moss, hardly thirty kilometres from Manchester, in the Peak District, to vast areas of one-time Scottish wetland, peatbogs have become so dry that streams and rivulets are finding it easy to cut through thousands of years of peat. They are washing the natural vegetation of the uplands down to the sea. Already, something around eight per cent of the Peak District National Park's moorland is bare, exposed peat. No acid bog has grown there for over a century. Because of a Government subsidy which encourages farmers to over-graze with sheep rather than to be conservationist, the fragile, nutrient-poor moorlands are carrying three times the numbers of sheep which sensible, sustainable farming used to allow. They are eating their way through the layer of living plants which protects the peat below.

In the case of forestry on old acid-bog sites, there is a largely false argument that since much of Scotland was once Caledonian pine forest, and further areas used, in the arboreal period (10,000 – 3,000 BC), to support growths of oak and birch, it is legitimate to put forestry back. The great problem with the afforestation argument is that the kind of forestry now planned and executed in the Highlands is much more demanding and intensive than anything which might have happened naturally. It is based on the premise, which no one has tested, that intensively grown trees on very fragile 'ground' like peat can be cropped regularly without destruction of the structure of the very growing medium

Above. The River Rea in Birmingham. On the left bank there has been no clearance work. The right bank has been devastated.

Right. The elimination of peat bog has been astonishingly thorough, as these maps of two regions testify.

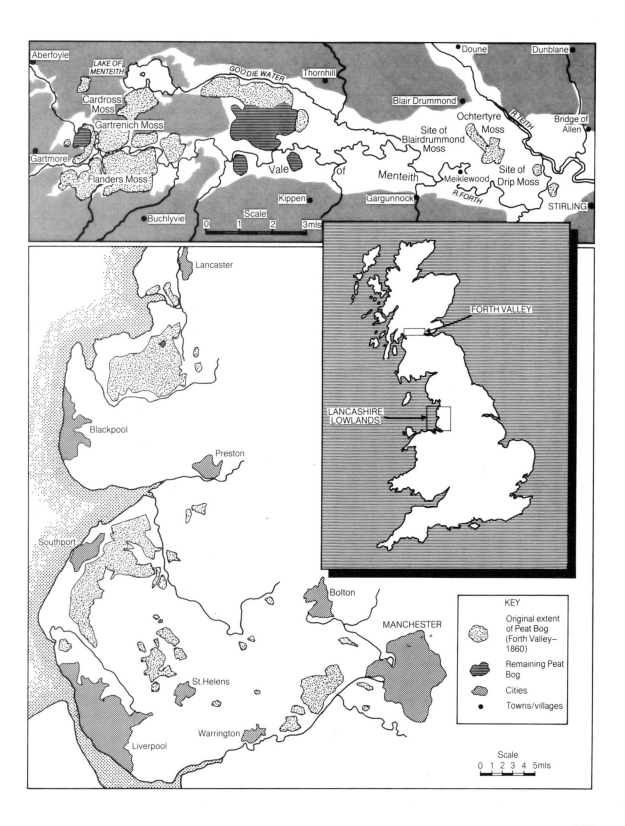

KEY

Original extent of Peat Bog (Forth Valley–1860)

Remaining Peat Bog

Cities

Towns/villages

in which they are planted.

Current forestry practice is, in effect, using much of Scotland as a giant grow-bag. It is entirely possible that after the first couple of 'crops' of wood fibre, the peat beneath the modern artificial forest will be entirely exhausted of its small and painstakingly accumulated nutritional value, and that no amount of fertilizers – even if they could be afforded with a low-profit crop like trees – would restore it. In that case, it is probable that much of Scotland's ground cover of peat will quite simply wash itself into the sea.

At least a previous generation could only work at the pace of nature. The old improved in-bye was certainly artificially induced, but it was achieved slowly enough, and largely using a system of encouraging two or three indigenous species at the expense of various others, to be economically and ecologically sustainable, if only enough fertilizer and lime were used. Modern farming and forestry are dangerous monocultures which pay little attention to the state of the soil and ground-cover they leave behind.

We have seen how the fenlands of East Anglia are nearing the end of their highly profitable life, due to the erosion of the peat of which they are composed. Soon very expensive measures will be required to maintain their potential as growing areas. In areas of one-time grazing marsh which have gone under the plough, farmers are often finding problems with unexpected acid chemical reactions from the deep soils, while many of the soils of ploughed marshland have been found to be notoriously unstable. These and other problems, not least the steep rises in petrochemical prices, will put many of the 'improvements' in doubt over the next few years.

It may well be that the many wise farmers who have been suspicious of the high-input, high-output farming of the post-war world will turn out not merely to have been good guardians of the nation's wildlife, but of her true farming interests too. Meanwhile, the tax-payer pays out £150 million annually in land drainage grants, of which a high proportion can be said to serve only one purpose: they make it profitable for farmers to switch from one heavily subsidized sort of farming, mostly dairy and beef, to another, arable, which is even more heavily subsidized. It is time for tax-payers, sensible farmers, and conservationists to see that their interests lie well together. Remember: we do not now have much time in which to save the last of our wild countryside.

This once-typical, unremarkable riverside scene is a good deal less common than once it was. Water Authorities have largely adopted a short-back-and-sides policy on many of their rivers, partly to keep themselves busy and partly to please the farmers, with whose interests they often, especially until recently, identified at the expense of seeing themselves as preserving habitat for people and wildlife.

Booklist

This is no more than a very personal selection of the books I found useful. Tansley is absolutely crucial and a delightful aid to understanding the British wildlife habitat. Mabey gives the most elegant and evocative account of why conservation matters and on what sorts of habitats it should be concentrated. Body is the sharpest and most stirring critic of the abuses of the farming industry.

Bellamy, David. *Botanic Man*, Hamlyn, 1978. Among the easiest-to-follow accounts of the world's botanical evolution and present.

Bellamy, David. *The Great Seasons*, Hodder and Stoughton, 1981. A vigorous account of the evolution of a Durham dale.

Body, R. *Agriculture: the Triumph and the Shame*, Maurice Temple Smith, 1982.

Forest and Wildlife Service of the Department of Lands (Irish Government) *Wetlands Discovered*. Price 35p from FWSDL, 22 Upper Merrion Street, Dublin 2. By the far the simplest and most elegant guide to the wetland habitat.

Geological Museum. *Britain Before Man*, HMSO, 1978. Excellent pamphlet on Britain's pre-history.

Geological Museum. *The Age of the Earth*, HMSO, 1980. Excellent pamphlet on the shaping of the planet.

Keble Martin, W. *The Concise British Flora in Colour*, Michael Joseph, 1974. Among the loveliest of the authoritative *floras*.

Mabey, R. *The Common Ground*, Arrow Books, 1981.

Newbold, C, Purseglove, J, and Holmes, N. *Nature Conservation and River Engineering*, £3.50 (incl P & P) from Nature Conservancy Council, see Useful Addresses.

Pennington, Winifred. *The History of British Vegetation*, The English University Press, 1974. A classic.

Phillips, Roger. *Grasses, Ferns, Mosses and Lichens*, Pan, 1980. Very clear and useful photographic guide.

Phillips, Roger. *Wild Flowers of Britain*, Pan, 1977. Very clear and useful photographic guide.

Ratcliffe, D A. *Highland Flora*, Highlands and Islands Development Board, Bridge House, Bank Street, Inverness, 1977.

Ratcliffe, D A, ed. for the Nature Conservancy Council. *A Nature Conservation Review*, Cambridge University Press, 1977. A wonderful guide to habitat gems in Britain. Volume One describes habitat types. Volume Two describes over 700 particular sites. Hideously expensive: a library job.

Shoard, M. *The Theft of the Countryside*, Maurice Temple Smith, 1980.

Tansley, A G. *Britain's Green Mantle*, Allen and Unwin, 1949. Beg, borrow, or steal this – but not mine.

Trueman, A E. *Geology and Scenery in England and Wales*, Pelican, 1980.

Vincent, J, and Lodge, G. *A Season of Birds, a Norfolk Diary, 1911*, Weidenfeld and Nicholson, 1980. Pretty; nostalgic.

Whitton, B. *Rivers, Lakes and Marshes*, Hodder and Stoughton, 1979. Excellent pocket guide to the flora and fauna of wetlands.

Useful Addresses

Council for the Protection of Rural England, 4 Hobart Place, London, SW1W OHY: members can join the Central Group (£5 annually) to support the national headquarters direct, or a county branch (varying fee). In any event, contact the London office for further information.

The Nature Conservancy Council, headquarters: 19/20 Belgrave Square, London, SW1X 8PY: a good range of wallcharts and explanatory leaflets: catalogue available from the NCC, Interpretative Branch, Attingham Park, Shrewsbury, SY4 4TW.

Royal Society for Nature Conservation, The Green, Nettleham, Lincs, LN2 2NR: supports County Naturalists' Trusts.

Royal Society for the Protection of Birds, The Lodge, Sandy, Beds, SG19 2DL: manages over seventy reserves in Britain and also works on overall habitat policy.

Friends of the Earth, 377 City Rd, London, EC1V 1NA: has a full-time wildlife and countryside campaigner.

picture acknowledgements

Most of the pictures in this book were taken by Glyn Satterley including the picture on the back of the cover. Others were taken as follows:
Zul Bhatia vii, 88, 96, 100/101, 144
BTA 34, 35
Stuart Housden (RSPB) 57, 60
Audrey Lincoln 137 (bottom)
Richard Lindsay 137 (top), 148, 156, 157, 161, 188/189
Richard North v
Picturepoint 28, 36/37, 72, 155
Press-tige 24/25, 32, 40/41, 45, 48/49, 61, 84/85, 109
Jeremy Purseglove 124, 125
Space Frontiers Ltd 17
Gwyn Williams (RSPB) 60, 68/69, 120 (top), 184, 185